VOLUME NINETY NINE

ADVANCES IN PROTEIN CHEMISTRY AND STRUCTURAL BIOLOGY

Peptide and Protein Vaccines

VOLUME NINETY NINE

ADVANCES IN PROTEIN CHEMISTRY AND STRUCTURAL BIOLOGY

Peptide and Protein Vaccines

Edited by

ROSSEN DONEV
Biomed Consult Ltd
Swansea, United Kingdom

ELSEVIER

AMSTERDAM • BOSTON • HEIDELBERG • LONDON
NEW YORK • OXFORD • PARIS • SAN DIEGO
SAN FRANCISCO • SINGAPORE • SYDNEY • TOKYO
Academic Press is an imprint of Elsevier

Academic Press is an imprint of Elsevier
225 Wyman Street, Waltham, MA 02451, USA
525 B Street, Suite 1800, San Diego, CA 92101-4495, USA
125 London Wall, London, EC2Y 5AS, UK
The Boulevard, Langford Lane, Kidlington, Oxford OX5 1GB, UK

First edition 2015

ISBN: 978-0-12-802827-8
ISSN: 1876-1623

For information on all Academic Press publications
visit our website at store.elsevier.com

CONTENTS

CONTRIBUTORS

Fernando Albericio
Catalysis and Peptide Research Unit, School of Health Sciences; School of Chemistry, Yachay Tech, Yachay City of Knowledge, Urcuqui, Ecuador; School of Chemistry and Physics, University of KwaZulu-Natal, Durban, South Africa; CIBER-BBN, Networking Centre on Bioengineering, Biomaterials and Nanomedicine, Barcelona Science Park; Department of Organic Chemistry, University of Barcelona, and Institute for Research in Biomedicine (IRB Barcelona), Barcelona, Spain

Beatriz Apellániz
Biophysics Unit (CSIC, UPV/EHU), and Department of Biochemistry and Molecular Biology, University of the Basque Country (UPV/EHU), Bilbao, Spain

Hsin-Wei Chen
Vaccine R&D Center, National Health Research Institutes, Miaoli, Taiwan

Prabhakar Cherkupally
Catalysis and Peptide Research Unit, School of Health Sciences, University of KwaZulu-Natal, Durban, South Africa

Soo Young Cho
Laboratory of Developmental Biology and Genomics, College of Veterinary Medicine, Research Institute for Veterinary Science BK21, Program for Veterinary Science, Seoul National University, Seoul, South Korea

Pele Chong
Vaccine R&D Center, National Health Research Institutes, Miaoli, Taiwan

Beatriz G. de la Torre
Catalysis and Peptide Research Unit, School of Health Sciences, University of KwaZulu-Natal, Durban, South Africa, and School of Chemistry, Yachay Tech, Yachay City of Knowledge, Urcuqui, Ecuador

Thavendran Govender
Catalysis and Peptide Research Unit, School of Health Sciences, University of KwaZulu-Natal, Durban, South Africa

Jui-Hsin Huang
Vaccine R&D Center, National Health Research Institutes, Miaoli, Taiwan

Zhi Jiang
Center for Systems Biology, School of Computer Science and Technology, and School of Medicine, Soochow University, Suzhou, 215123 PR China

Dong Seok Kim
Research and Development Center, Peptron, Inc., Daejeon, South Korea

Hendrik G. Kruger
Catalysis and Peptide Research Unit, School of Health Sciences, University of KwaZulu-Natal, Durban, South Africa

Hyung-Joo Kwon
Department of Microbiology, and Center for Medical Science Research, College of Medicine, Hallym University, Chuncheon, South Korea

Young Seek Lee
Division of Molecular and Life Sciences, College of Science and Technology, Hanyang University, Ansan, South Korea

Younghee Lee
Department of Biochemistry, College of Natural Sciences, Chungbuk National University, Cheongju, South Korea

Chih-Hsiang Leng
Vaccine R&D Center, National Health Research Institutes, Miaoli, Taiwan

Shih-Jen Liu
Vaccine R&D Center, National Health Research Institutes, Miaoli, Taiwan

José L. Nieva
Biophysics Unit (CSIC, UPV/EHU), and Department of Biochemistry and Molecular Biology, University of the Basque Country (UPV/EHU), Bilbao, Spain

Suhas Ramesh
Catalysis and Peptide Research Unit, School of Health Sciences, University of KwaZulu-Natal, Durban, South Africa

Bairong Shen
Center for Systems Biology, School of Computer Science and Technology, Soochow University, Suzhou, 215006 PR China

Ke Shen
Center for Systems Biology, School of Computer Science and Technology, Soochow University, Suzhou, 215006 PR China

Li Shen
Center for Systems Biology, School of Computer Science and Technology, Soochow University, Suzhou, 215006 PR China, and Institute of Biological Sciences and Biotechnology, Donghua University, Shanghai, 201620 PR China

Jing Wang
Center for Systems Biology, School of Computer Science and Technology, Soochow University, Suzhou, 215006 PR China

Hyun Yang
Research and Development Center, Peptron, Inc., Daejeon, South Korea

PREFACE

Conventional vaccine strategies proved highly efficacious for decades in reducing mortality and morbidity due to infectious diseases. The major drawbacks of conventional vaccines, such as those including whole organisms or large proteins, seem to be the inclusion of unwanted antigens that not only contributes little to the protective immune response but also are likely to promote allergenic and/or reactogenic responses. Thus, peptide vaccines are an attractive alternative strategy based on the engineering of short peptide fragments that induce highly targeted immune responses, allowing the avoidance of allergenic and/or reactogenic reactions. A successful vaccine candidate should meet several criteria: immunogenicity, specificity, protective activity, and durability. First, it should be immunogenic, i.e., it should be able to elicit an immune response, both humoral and cell-mediated, that leads to the blocking and eradication of the disease-causing agent and clearance of the affected cells. Second, this immune response must specifically target (bind with a certain affinity) the region on the pathogen that has been mimicked by the immunogen used as a vaccine, that being a peptide, a protein, or a whole organism, and it must avoid cross-reaction with other self-antigens to prevent autoimmunity. Third, the elicited immune response should be able to prevent the establishment of a disease if the aim is to pursue a prophylactic vaccine, which is usually the case in pathogen-mediated infections, including prions. Finally, the vaccine must be able to induce the production of B and T memory cells, both required to ensure re-elicitation of the protective immune response should the organism encounter the pathogenic agent even years after immunization (Apellániz & Nieva, 2015).

A peptide vaccine usually consists of one or more peptide sequences of more than 15 amino acids long that induce B and T cell stimulation when presented by itself or bound to carrier proteins, scaffolds, or supramolecular complexes (e.g., liposomes). Moreover, peptides from one or several strains of the same pathogen might be included to ensure proper coverage of pathogen variability. Some peptide-based formulations devised following these strategies have rendered vaccines effective in preventing viral infection in animals (Bittle et al., 1982; Langeveld et al., 1994).

First chapter in this volume discusses new promising strategies of peptide vaccine development recently progressed in preclinical and/or clinical stage with main focus on the roles of peptides in the vaccine formulation from epitope to adjuvant. Second article in this thematic volume gives an overview of applications of lipid vesicles (liposomes) to the development of membrane-proximal external region-targeting vaccines, both as type B adjuvants and epitope structure-shaping devices. This chapter introduces a new paradigm in peptide vaccine development: the structural stabilization of peptide epitopes through contacts with the membrane surface. The third chapter in this volume discusses strategies for developing successful prophylactic and therapeutic vaccines using lipoprotein-based immunogens that are safe, cost-effective, and suitable for human use. Authors support their point of view with a number of examples that demonstrate the merit of lipoproteins with intrinsic adjuvant properties for novel vaccine development. In the fourth chapter, authors review a strategy for improving the efficacy of peptide vaccines. They discuss recent studies providing a potent method of epitope screening and antibody production without conventional carriers. Instead, they adopted Lipoplex(O), comprising a natural phosphodiester bond CpG-DNA and a specific liposome complex, as an adjuvant. Lipoplex(O) induces potent stimulatory activity in humans and mice, and immunization of mice with several peptides co-encapsulated with Lipoplex(O) without carriers significantly induces each peptide-specific IgG2a production. This strategy can be applied in development of therapeutic antibodies or in defense against pandemic infectious diseases through rapid screening of potent B cell epitopes. In the fifth article of the thematic volume, authors discuss several different chemistries that have been pursued to obtain novel platforms onto which antigenic epitopes can be tethered, with the aim to achieve a higher antibody response. In this regard, they review the chemical strategies developed for the presentation of peptide epitopes. The final sixth chapter in this volume focuses on the role of mutations in viral proteins for the design strategy of vaccines against the viruses, which has been exemplified in hepatitis B virus.

The aim of this volume is to promote further research and development in the design of peptide vaccines in order to achieve highly targeted immune responses against different pathological conditions while avoiding allergenic and/or reactogenic reactions.

Dr. Rossen Donev
Biomed Consult Ltd.

REFERENCES

Apellániz, B., & Nieva, J. L. (2015). The use of liposomes to shape epitope structure and modulate immunogenic responses of peptide vaccines against HIV MPER. *Advances in Protein Chemistry and Structural Biology*, *99*, 15–54.

Bittle, J. L., Houghten, R. A., Alexander, H., Shinnick, T. M., Sutcliffe, J. G., Lerner, R. A., et al. (1982). Protection against foot-and-mouth disease by immunization with a chemically synthesized peptide predicted from the viral nucleotide sequence. *Nature*, *298*(5869), 30–33.

Langeveld, J. P., Casal, J. I., Cortes, E., van de Wetering, G., Boshuizen, R. S., Schaaper, W. M., et al. (1994). Effective induction of neutralizing antibodies with the amino terminus of VP2 of canine parvovirus as a synthetic peptide. *Vaccine*, *12*(15), 1473–1480.

CHAPTER ONE

Peptide Immunotherapy in Vaccine Development: From Epitope to Adjuvant

Hyun Yang, Dong Seok Kim[1]
Research and Development Center, Peptron, Inc., Daejeon, South Korea
[1]Corresponding author: e-mail address: dskim@peptron.co.kr

Contents

Abstract

Vaccines are designed to educate the host immune system to prevent infectious disease or to fight against various diseases such as cancers. Peptides were first employed to provide specific immune responses while minimizing unintended allergenic or reactogenic adverse effects. Discoveries of virus or cancer-specific antigens and the advanced knowledge of immunology accelerate the peptide vaccine development. Despite the overwhelming research pipelines, a very few of them reached to market approvals or phase III clinical trials, because of the lack of efficacy. Several strategies for the next generation peptide vaccines are devised to overcome the weak immunogenicity and the poor delivery. In this review, we discuss the new promising strategies of peptide vaccine development which are recently developed in preclinical and/or clinical stage focusing the roles of peptides in the vaccine formulation from epitope to adjuvant. Additionally, we discuss the future perspectives of peptide vaccine and immunotherapy.

1. INTRODUCTION

Synthetic peptide vaccines are usually composed of 20–30 amino acids containing the specific epitope of an antigen related to infectious and/or chronic diseases including cancers. Peptide vaccines theoretically have several advantages over other types of vaccines such as conventional vaccines

Advances in Protein Chemistry and Structural Biology, Volume 99
ISSN 1876-1623
http://dx.doi.org/10.1016/bs.apcsb.2015.03.001

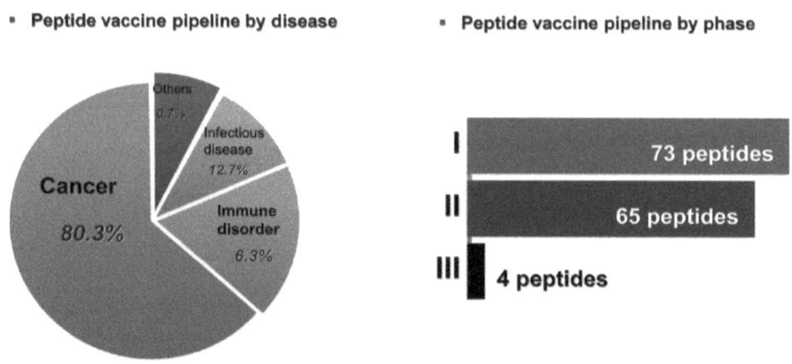

Figure 1 Current development of peptide vaccine (2015).

and newly developed DNA or cellular vaccines (Ingolotti, Kawalekar, Shedlock, Muthumani, & Weiner, 2010). Easy synthesis with low cost, increased stability, and relative safety are generally demonstrated in numerous preclinical and clinical studies. In addition, peptide vaccines have no limitation in target diseases from virus infection to Alzheimer disease and even allergy (Fig. 1A) (Larche, 2007; Mocellin, Pilati, & Nitti, 2009; Nava-Parada, Forni, Knutson, Pease, & Celis, 2007; Park et al., 2014). Peptide vaccines can be designed with self- or nonself-antigen to properly balance the immune responses, which is not possible for conventional vaccines (Purcell, McCluskey, & Rossjohn, 2007). However, it was recently reported that no peptide vaccine is approved by FDA, although more than 500 peptides had progressed to clinical trials (Li, Joshi, Singhania, Ramsey, & Murthy, 2014). According to ClinicalTrials.gov, a public database which is a service of the U.S. National Institute of Health, there are 73 clinical trials in phase I, 65 clinical trials in phase II, and 4 clinical trials in phase III in the search result of peptide vaccine in March 2015 (Fig. 1B). The series of failures in peptide vaccines in clinical trials suggest several issues critical for the successful development of peptide vaccines. These include (1) limitation of single peptide epitopes as vaccine candidates, (2) an immune evasion, (3) the failure to elicit the controlled and prolonged immune response, (4) a lack of efficacy, and (5) the inappropriate design of clinical trials. In this review, we discuss the roles of peptides in vaccine formulation focusing on the innovative approach overcoming those limitations in the recent clinical studies or the researches close to clinical development.

2. CANCER VACCINE

The rational of cancer vaccine development is that tumor cells can be eradicated by induction of cytotoxic T lymphocyte (CTL) response against

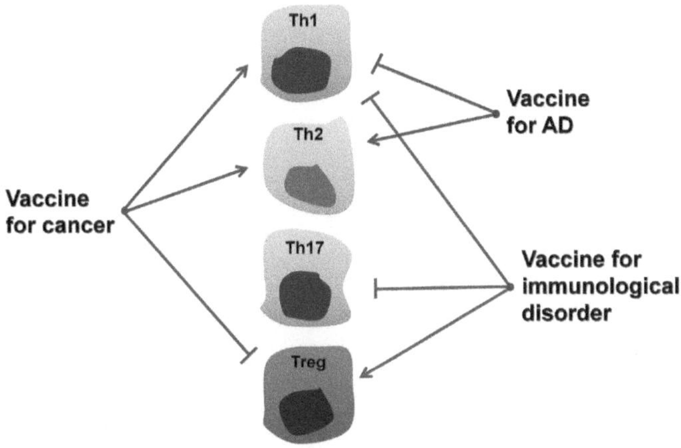

Figure 2 Rational design of peptide vaccines for specific diseases.

tumor-associated antigen (TAA) (Fig. 2) (Inoda et al., 2011; Melief & van der Burg, 2008). TAAs are generally processed in antigen-presenting cells (APCs) and presented to T cells in a human leukocyte antigen (HLA)-restricted pathway (Hirano et al., 2006; Parmiani et al., 2002). TAA-specific CTL can be activated and attack cancer cells recognized by TAA expression (Parmiani et al., 2002). Many peptide cancer vaccines are designed to stimulate T cells but most of them failed to show clinical benefits in clinical trials even though some of the vaccines successfully activated APC and TAA-specific T cells. The lack of vaccine efficacy was thought to result from HLA restriction, diversity of cancer phenotype, and immune evasion (Chentoufi et al., 2010; Khong & Restifo, 2002). To overcome these limitations, multiple peptides are employed in the development of cancer vaccine formulation rather than a single peptide.

IMA901 (Immatics) is a peptide cancer vaccine composed of multiple tumor-associated peptides (TUMAPs) using GM-CSF as adjuvant (Walter, Weinschenk, Reinhardt, & Singh-Jasuja, 2013; Walter et al., 2012). TUMAPs consist of 10 different peptide epitopes (Fig. 3), which are found to be overexpressed in the majority of renal cell carcinoma (RCC) (Bedke & Stenzl, 2013; Rausch, Kruck, Stenzl, & Bedke, 2014).

In a phase II clinical trial of IMA901, a randomized trial with 68 HLA-A*02-positive RCC patients, the group 1 patients (n=35) received IMA901 while the group 2 patients (n=33) received IMA901 with the pretreatment of cyclophosphamide (300 mg/m, n=35) or IMA901 preceded by a single immune modulatory dose of i.v. cyclophosphamide (group 2 [n=33]; 300 mg/m^2) (Pal, Hu, & Figlin, 2013; Walter et al., 2012). The patient

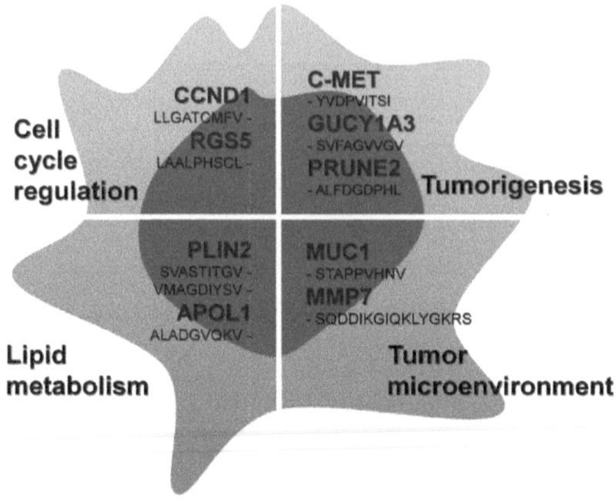

Figure 3 Tumor-associated peptides (TUMAPs) in IMA901 cancer vaccine.

groups then received up to 17 injections of IMA901 plus GM-CSF over a period of up to 9 months (Walter et al., 2012). Among those patients that elicited an immune response to IMA901, cyclophosphamide pretreatment was associated with a significantly prolonged survival (Rausch et al., 2014; Walter et al., 2012). However, cyclophosphamide pretreatment had no impact on survival in the subset of patients who lacked an immune response. IMA901 is currently in a phase III clinical trial and has been granted orphan drug designation in the United States and Europe for the treatment RCC in HLA-A*02-positive patients.

On the other hand, GV1001 (KAEL-GemVax) is a cancer vaccine composed of a single synthetic peptide containing multiple epitopes in a 16-amino acid-long peptide. The peptide is derived from the human telomerase reverse transcriptase (Brunsvig et al., 2006; Park et al., 2014). Because GV1001 could bind to multiple HLA class I and HLA class II molecules (Kyte, 2009), it may therefore elicit combined CD4/CD8 T cell response, considered important to initiate tumor eradication and also long-term memory (Raval, Sharabi, Walker, Drake, & Sharma, 2014). In a phase II study with advanced pancreatic cancer patients, GV1001 showed total immune responses (63%) and a greater median survival (216 days; 146–323) in immune responders than in nonresponders (88 days; 53–190) (Bernhardt et al., 2006; Middleton et al., 2014). However, in a three-group, open-label, randomized phase III trial with locally advanced or metastatic pancreatic cancer patients, GV1001 fails to improve overall survival. Interestingly, cytokine analysis for the prediction of biomarker revealed that in the subset of patients with high eotaxin (CLL11) level GV1001 improved the median overall survival (high eotaxin = 14.8 vs. low eotaxin = 7.9) (Neoptolemos, 2014). Although the efficacy of GV1001 depending on eotaxin levels should be examined in a large phase III clinical trial, GV1001 recently received new drug approval in Korea for the pancreatic cancer patients with high level of eotaxin.

3. ALLERGY VACCINE

In normal physiological condition, body's immune system operates in steady-state condition to maintain immune homeostasis (Liu et al., 2007). However, abnormal failure of this homeostasis results in immune disorders such as autoimmune diseases and allergies (Burton & Oettgen, 2011). As we more understand the immunopathological events in immune-related diseases through the series of clinical studies, many innovative immunotherapies have been designed including vaccines.

Cat-Pad (Circassia) is a cat allergy vaccine composed of seven synthetic peptide immunogen-regulatory epitopes originated from the major cat allergen Fel d 1 (Worm, Patel, & Creticos, 2013). Because native allergen or long peptides can induce allergenic responses, Cat-Pad employed seven short peptides (13–17 amino acids) derived from binding analysis to common HLA-DR molecules (Table 1) which enable the peptides to bind to broad range of HLA molecules (Worm et al., 2013). In the randomized, double-blind, placebo-controlled study (phase IIb), the vaccine was well tolerated and demonstrated clinical benefits in both Total Rhinoconjunctivitis

Table 1 Fel d 1 Peptide Sequences in Cat-Pad, a Cat Allergy Vaccine

Peptide Origin	Length	Epitope	Sequence
Fel d 1 chain 1 (3–15)	13	DR3,15	CPAVKRDVDLFLT
Fel d 1 chain 1 (23–38)	16	DR1,4,11,13,15	EQVAQYKALPVVLENA
Fel d 1 chain 1 (29–45)	17	DR1,3,4,11,13,15 DRB4,5	KALPVVLENARILNCV
Fel d 1 chain 1 (39–55)	17	DR3,4	RILKNCVDAKMTEEDKE
Fel d 1 chain 1 (54–69)	16	DR11,13,15	KENALSLLDKIYTSPL
Fel d 1 chain 2 (40–55)	16	DR15 DRB4	TAMKKIQDCYVENGLI
Fel d 1 chain 2 (56–71)	16	DR4,11 DRB4	SRVLDGLVMTTISSSK

Symptom Scores (TRSS) and Total Nasal Symptom Scores (TNSS) that persisted 2 years after treatment (Aydogan et al., 2013; Patel et al., 2013). The randomized, double-blind, placebo-controlled, multicenter, international field study (phase III) is currently on-going for patients with cat allergen-induced rhinoconjunctivitis.

AllerT (Anergis) is a birch pollen allergy vaccine composed of eight contiguous overlapping peptides (COPs) derived from Bet v I allergen (Pellaton et al., 2013). COPs are 40–80-amino acid-long synthetic peptides (Fig. 4) that cover most T cell epitopes in the compete allergen (Pellaton et al., 2013). Because the cross-reactivity with patient's IgE is responsible for the major adverse effect in the immunotherapy of allergies, COPs are designed to encompass T cell epitopes but to disrupt IgE conformational epitopes. In a placebo-controlled, double-blind, randomized multicenter trial (phase II), AllerT was well tolerated and demonstrated significant improvements in the combined Rhinoconjunctivitis Symptom and Medication Score (RSMS), in the total score of the Rhinoconjunctivitis Quality of Life Questionnaire (Mini-RQLQ), and in the Rhinoconjunctivitis Symptom Score (Pellaton et al., 2013). These result suggested that the efficacy of AllerT could be tested in the advanced phase III clinical trial.

4. ALZHEIMER'S DISEASE VACCINE

Neurodegenerative diseases such as Alzheimer's disease and Parkinson's disease are characterized by the aggregation of misfolded or toxic proteins that are thought to play a critical role in disease progression (Maiti, Manna, Veleri, & Frautschy, 2014; Serrano-Pozo, Frosch, Masliah, & Hyman, 2011). Especially, Alzheimer's disease pathogenesis is generally explained by the toxicity of Aβ accumulation and deposition that lead to neuronal cell death and dysfunction causing dementia at the end (Serrano-Pozo et al., 2011). According to this pathological mechanism of AD, active and passive immunotherapy targeting Aβ aggregation was rationally designed including the anti-AD peptide vaccine, AN1792 (Gilman et al., 2005; Patton et al., 2006). However, the AN1792 clinical trial with AD patient was stopped because of the adverse effect, encephalitis resulting from inflammatory T cell activation. It turned out later that the adjuvant Qs-21 caused Th-1 immune response against Aβ peptide.

To avoid this unwanted immune response, several promising alternatives were developed by using modified Aβ peptide, ACC-001 (Pfizer), CAD106 (Novartis), and Affitope AD02 (Affiris AG). ACC-001 uses only B cell

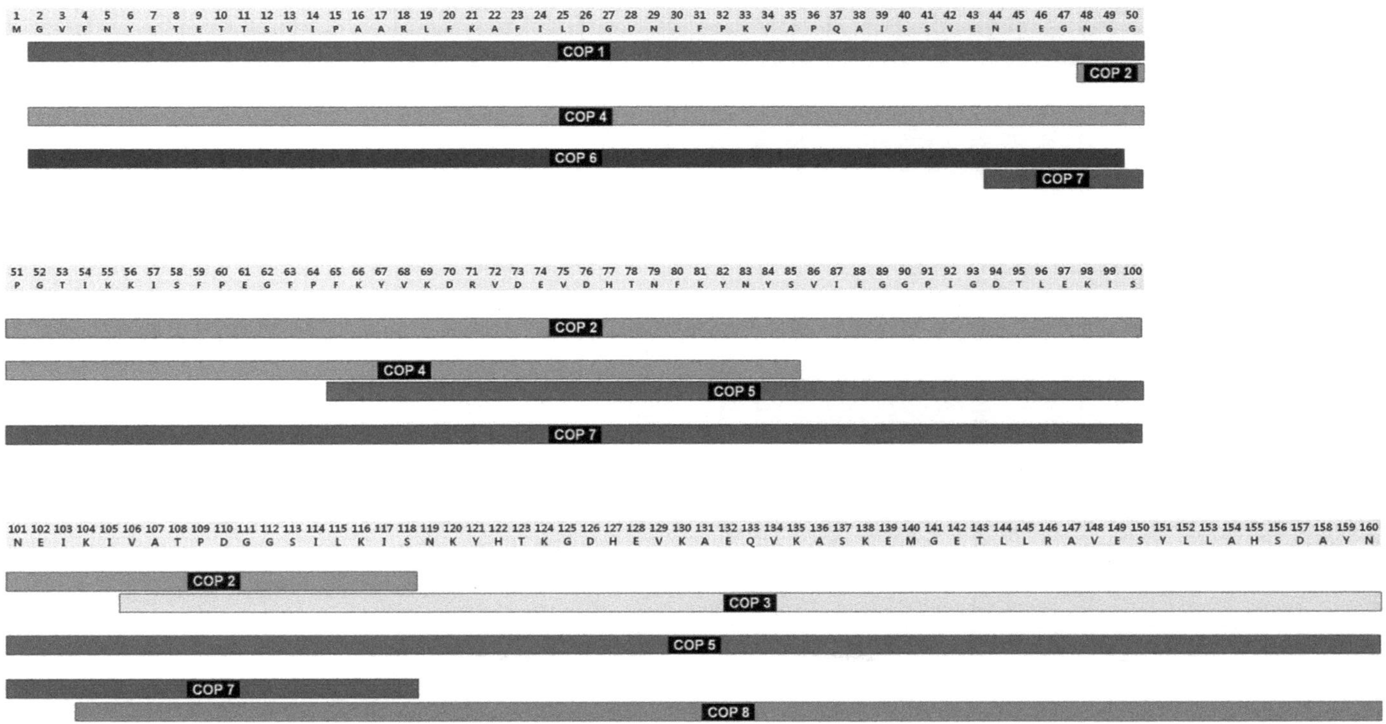

Figure 4 Contiguous overlapping peptides (COPs) from Bet v I allergen.

epitope, Aβ1-6 with adjuvant Qs21 instead of Aβ42 containing the T cell epitopes (Li, Liu, Wang, & Jiang, 2013; Panza et al., 2014). CAD106 also uses B cell epitope, Aβ1-6 that is coupled to carrier protein (Lambracht-Washington & Rosenberg, 2013). Affitope employed peptide mimicking Aβ1-6 epitope instead of Aβ1-6 itself. As the lessons from AN1792 clinical study brought in innovative vaccine strategies (Lambracht-Washington & Rosenberg, 2013), the results of current phase II clinical trials of these vaccines will pave the way to the next step of the anti-AD vaccine development.

Recently, another peptide vaccine entered phase Ib clinical trial for AD, targeting phospho-Tau protein that is another hallmark of AD pathology, besides Aβ plaques (Sigurdsson, 2008). ACI-35 (AC immune/Janssen) contains 16-amino acid-long peptide from human Tau protein (393–408) phosphorylated on Ser396 and Ser404 (Evans et al., 2000; Mondragon-Rodriguez, Perry, Luna-Munoz, Acevedo-Aquino, & Williams, 2014). In animal study using transgenic Tau.P301L mice, ACI-35 was able to induce specific antibodies against phosphorylated Tau, which lead to a reduction of pathogenic Tau (Theunis et al., 2013). Because Tau and Aβ are two key factors in pathogenesis of AD, it is worth to test the combination vaccine targeting both targets simultaneously.

5. ADJUVANT AND DELIVERY

Most peptide vaccines failed in clinical trials because of the lack of efficacy (Pilon et al., 2013; Rezvani et al., 2011). Development of innovative adjuvant and delivery methods is required to get enough stimulation of appropriate and prolonged immune response (Petrovsky & Aguilar, 2004). Recently, a breakthrough immunotherapy for cancer treatment is developed by targeting immune costimulatory receptors such as CTLA-4 and PD-1 (Ott, Hodi, & Robert, 2013; Parry et al., 2005). The same strategy can be adopted in vaccine development. Although it is not a synthetic peptide, one of the best examples is the PROSTVAC (Bavarian Nordic). PROSTVAC employs the poxviral vector that contains four genes encoding prostate-specific antigen as well as three costimulatory molecules (TRICOM), B7-1, ICAM-1, and LFA (DiPaola et al., 2006). In a randomized placebo-controlled phase II vaccine trial, the treatment of PROSTVAC improved median survival benefit by 8.5 months ($p=0.006$) improved by 8.5 months and the death rate reduced by 44% (Gulley, Madan, & Schlom, 2011). PROSTVAC is currently in a pivotal phase III clinical study.

The Pan DR epitope peptide (PADRE) is another class of adjuvant which can be chemically synthesized with peptide epitope such as tumor-specific antigen (Alexander et al., 2000). PADRE is designed to overcome the problems resulting from the polymorphism of HLA-DR in the human population and can generate antigen-specific CD4 + T cells that bind various MHC class II molecules with high affinity (La Rosa et al., 2012). Because PADRE has been extensively tested in experimental animal models and shown to be safe in several human studies, further clinical development of PADRE with diverse peptide epitopes are expected in near future.

Two interesting peptides modulating immune responses were identified and in development in human clinical trial. One is helper/killer-hybrid epitope long peptide and the other is Tregitope (Gershwin & Shoenfeld, 2013).

Helper/killer-hybrid epitope long peptide (Hokkaido University) is an artificial 40-amino acid-long peptides containing both helper and killer epitopes designed to induce superior Th1 immune responses (Takahashi et al., 2012). In a phase I clinical trial of the Survivin-H/K-HELP vaccine, the vaccine induced Survivin-peptide specific cellular and humoral responses and exhibits a complete response in a patient diagnosed with postoperative chemo-resistant, radio-resistant lateral deep cervical node recurrence of breast cancer (Ohtake et al., 2014).

On the other hand, Tregitopes (Epivax) can bind to multiple MHC II molecules and activate regulatory immune responses. Originally, Tregitopes are peptide epitopes located in immunoglobulin and are able to induce regulatory T cell-mediated immune suppression (Su, Rossi, De Groot, & Scott, 2013). Tregitopes are expected to have unique position in peptide vaccine development targeting autoimmune disease such as type 1 diabetes mellitus (Cousens et al., 2013). Immunosuppressive effects of Tregitopes were already proven in diverse animal autoimmune disease models whether Tregitopes are administered alone or coadministered with target antigens (Cousens et al., 2013; Su et al., 2013). Clinical proof-of-concept research is essential for the further development of Tregitopes as well as the precise understanding of the immunosuppressive mechanism.

One of the critical limitations of peptide vaccines is the absence of effective delivery system (Brooks, Pouniotis, Tang, Apostolopoulos, & Pietersz, 2010; Li et al., 2014). Peptides are easy to be degraded by lots of enzymes *in vivo* so that only a small portion of the injected peptides can be taken up by APCs and be processed by the endosomal antigen-processing pathway (Erazo-Oliveras, Muthukrishnan, Baker, Wang, & Pellois, 2012; Nakase, Kobayashi, & Futaki, 2010). To facilitate the uptake of peptides by APCs, a delivery system for peptide-based vaccines based on a polyacrylate star polymer conjugated with target-specific peptide epitope. In a recent study with the polymer–peptide conjugate S4-8Qmin containing HPV-16 E7 protein epitope (Liu et al., 2015), the polymer–peptide conjugate was efficiently reached to APCs and induced cytotoxic T cell immune response against cancer growth in animal TC-1 tumor model (Nguyen et al., 2013). Although it is still in a very early stage, the peptide star polymer can be a promising candidate for peptide vaccine delivery platform technology without additional use of other adjuvants.

6. PERSPECTIVES AND CONCLUSIONS

Series of failures in peptide vaccine clinical trials, though only a few vaccines are approved yet, lay the groundwork for the era of immunotherapy. Deep understanding of human immunology guides the following vaccine studies to develop rational approaches such as design of epitope combination, adjuvant for safe and appropriate immune response, efficient delivery, and clinical study design. More than 400 TAAs are already validated and ready to be used in combination for cancer vaccine formulation (Buonaguro, Petrizzo, Tornesello, & Buonaguro, 2011). Recently developed immune check point inhibitors targeting PD-1/PD-L1 or CTLA-4 can be employed as adjuvants for cancer vaccine. Microparticles are one of the choices for the delivery of peptides by their number, characteristics, and combination of other functional ingredients. Most importantly, proper design of clinical study will be the key for the success: dosage regimens, combination of chemotherapy or other immunotherapy, development of biomarkers, patient selection, specific induction or inhibition of regulatory immune response, etc.

Moreover, cellular and molecular studies on patient with vaccine responses lead to development of personalized immune therapies such as adoptive cell transfer and chimeric antigen receptor T cell therapies. Universal cell therapies are also expected when it combined with gene-editing technology such as CRISPR in near future.

REFERENCES

Alexander, J., del Guercio, M. F., Maewal, A., Qiao, L., Fikes, J., Chesnut, R. W., et al. (2000). Linear PADRE T helper epitope and carbohydrate B cell epitope conjugates induce specific high titer IgG antibody responses. *Journal of Immunology*, *164*, 1625–1633.

Aydogan, M., Eifan, A. O., Keles, S., Akkoc, T., Nursoy, M. A., Bahceciler, N. N., et al. (2013). Sublingual immunotherapy in children with allergic rhinoconjunctivitis mono-sensitized to house-dust-mites: A double-blind-placebo-controlled randomised trial. *Respiratory Medicine, 107*, 1322–1329.

Bedke, J., & Stenzl, A. (2013). IMA901: A peptide vaccine in renal cell carcinoma. *Expert Opinion on Investigational Drugs, 22*, 1329–1336.

Bernhardt, S. L., Gjertsen, M. K., Trachsel, S., Moller, M., Eriksen, J. A., Meo, M., et al. (2006). Telomerase peptide vaccination of patients with non-resectable pancreatic cancer: A dose escalating phase I/II study. *British Journal of Cancer, 95*, 1474–1482.

Brooks, N. A., Pouniotis, D. S., Tang, C. K., Apostolopoulos, V., & Pietersz, G. A. (2010). Cell-penetrating peptides: Application in vaccine delivery. *Biochimica et Biophysica Acta, 1805*, 25–34.

Brunsvig, P. F., Aamdal, S., Gjertsen, M. K., Kvalheim, G., Markowski-Grimsrud, C. J., Sve, I., et al. (2006). Telomerase peptide vaccination: A phase I/II study in patients with non-small cell lung cancer. *Cancer Immunology, Immunotherapy, 55*, 1553–1564.

Buonaguro, L., Petrizzo, A., Tornesello, M. L., & Buonaguro, F. M. (2011). Translating tumor antigens into cancer vaccines. *Clinical and Vaccine Immunology, 18*, 23–34.

Burton, O. T., & Oettgen, H. C. (2011). Beyond immediate hypersensitivity: Evolving roles for IgE antibodies in immune homeostasis and allergic diseases. *Immunological Reviews, 242*, 128–143.

Chentoufi, A. A., Dasgupta, G., Christensen, N. D., Hu, J., Choudhury, Z. S., Azeem, A., et al. (2010). A novel HLA (HLA-A*0201) transgenic rabbit model for preclinical evaluation of human CD8+ T cell epitope-based vaccines against ocular herpes. *Journal of Immunology, 184*, 2561–2571.

Cousens, L. P., Su, Y., McClaine, E., Li, X., Terry, F., Smith, R., et al. (2013). Application of IgG-derived natural Treg epitopes (IgG Tregitopes) to antigen-specific tolerance induction in a murine model of type 1 diabetes. *Journal of Diabetes Research, 2013*, 621693.

DiPaola, R. S., Plante, M., Kaufman, H., Petrylak, D. P., Israeli, R., Lattime, E., et al. (2006). A phase I trial of pox PSA vaccines (PROSTVAC-VF) with B7-1, ICAM-1, and LFA-3 co-stimulatory molecules (TRICOM) in patients with prostate cancer. *Journal of Translational Medicine, 4*, 1.

Erazo-Oliveras, A., Muthukrishnan, N., Baker, R., Wang, T. Y., & Pellois, J. P. (2012). Improving the endosomal escape of cell-penetrating peptides and their cargos: Strategies and challenges. *Pharmaceuticals (Basel), 5*, 1177–1209.

Evans, D. B., Rank, K. B., Bhattacharya, K., Thomsen, D. R., Gurney, M. E., & Sharma, S. K. (2000). Tau phosphorylation at serine 396 and serine 404 by human recombinant tau protein kinase II inhibits tau's ability to promote microtubule assembly. *The Journal of Biological Chemistry, 275*, 24977–24983.

Gershwin, M. E., & Shoenfeld, Y. (2013). Abul Abbas: An epitome of scholarship. *Journal of Autoimmunity, 45*, 1–6.

Gilman, S., Koller, M., Black, R. S., Jenkins, L., Griffith, S. G., Fox, N. C., et al. (2005). Clinical effects of Abeta immunization (AN1792) in patients with AD in an interrupted trial. *Neurology, 64*, 1553–1562.

Gulley, J. L., Madan, R. A., & Schlom, J. (2011). Impact of tumour volume on the potential efficacy of therapeutic vaccines. *Current Oncology, 18*, e150–e157.

Hirano, N., Butler, M. O., Xia, Z., Berezovskaya, A., Murray, A. P., Ansen, S., et al. (2006). Efficient presentation of naturally processed HLA class I peptides by artificial antigen-presenting cells for the generation of effective antitumor responses. *Clinical Cancer Research, 12*, 2967–2975.

Ingolotti, M., Kawalekar, O., Shedlock, D. J., Muthumani, K., & Weiner, D. B. (2010). DNA vaccines for targeting bacterial infections. *Expert Review of Vaccines, 9*, 747–763.

Inoda, S., Hirohashi, Y., Torigoe, T., Morita, R., Takahashi, A., Asanuma, H., et al. (2011). Cytotoxic T lymphocytes efficiently recognize human colon cancer stem-like cells. *The American Journal of Pathology, 178*, 1805–1813.

Khong, H. T., & Restifo, N. P. (2002). Natural selection of tumor variants in the generation of "tumor escape" phenotypes. *Nature Immunology*, *3*, 999–1005.

Kyte, J. A. (2009). Cancer vaccination with telomerase peptide GV1001. *Expert Opinion on Investigational Drugs*, *18*, 687–694.

Lambracht-Washington, D., & Rosenberg, R. N. (2013). Advances in the development of vaccines for Alzheimer's disease. *Discovery Medicine*, *15*, 319–326.

Larche, M. (2007). Immunotherapy with allergen peptides. *Allergy, Asthma and Clinical Immunology*, *3*, 53–59.

La Rosa, C., Longmate, J., Lacey, S. F., Kaltcheva, T., Sharan, R., Marsano, D., et al. (2012). Clinical evaluation of safety and immunogenicity of PADRE-cytomegalovirus (CMV) and tetanus-CMV fusion peptide vaccines with or without PF03512676 adjuvant. *The Journal of Infectious Diseases*, *205*, 1294–1304.

Li, W., Joshi, M. D., Singhania, S., Ramsey, K. H., & Murthy, A. K. (2014). Peptide vaccine: Progress and challenges. *Vaccine*, *2*, 515–536.

Li, Y., Liu, Y., Wang, Z., & Jiang, Y. (2013). Clinical trials of amyloid-based immunotherapy for Alzheimer's disease: End of beginning or beginning of end? *Expert Opinion on Biological Therapy*, *13*, 1515–1522.

Liu, T. Y., Giddam, A. K., Hussein, W. M., Jia, Z., McMillan, N. A., Monteiro, M. J., et al. (2015). Self-adjuvanting therapeutic peptide-based vaccine induce CD8+ cytotoxic T lymphocyte responses in a murine human papillomavirus tumor model. *Current Drug Delivery*, *12*, 3–8.

Liu, Y. J., Soumelis, V., Watanabe, N., Ito, T., Wang, Y. H., Malefyt Rde, W., et al. (2007). TSLP: An epithelial cell cytokine that regulates T cell differentiation by conditioning dendritic cell maturation. *Annual Review of Immunology*, *25*, 193–219.

Maiti, P., Manna, J., Veleri, S., & Frautschy, S. (2014). Molecular chaperone dysfunction in neurodegenerative diseases and effects of curcumin. *BioMed Research International*, *2014*, 495091.

Melief, C. J., & van der Burg, S. H. (2008). Immunotherapy of established (pre)malignant disease by synthetic long peptide vaccines. *Nature Reviews. Cancer*, *8*, 351–360.

Middleton, G., Silcocks, P., Cox, T., Valle, J., Wadsley, J., Propper, D., et al. (2014). Gemcitabine and capecitabine with or without telomerase peptide vaccine GV1001 in patients with locally advanced or metastatic pancreatic cancer (TeloVac): An open-label, randomised, phase 3 trial. *Lancet Oncology*, *15*, 829–840.

Mocellin, S., Pilati, P., & Nitti, D. (2009). Peptide-based anticancer vaccines: Recent advances and future perspectives. *Current Medicinal Chemistry*, *16*, 4779–4796.

Mondragon-Rodriguez, S., Perry, G., Luna-Munoz, J., Acevedo-Aquino, M. C., & Williams, S. (2014). Phosphorylation of tau protein at sites Ser(396–404) is one of the earliest events in Alzheimer's disease and Down syndrome. *Neuropathology and Applied Neurobiology*, *40*, 121–135.

Nakase, I., Kobayashi, S., & Futaki, S. (2010). Endosome-disruptive peptides for improving cytosolic delivery of bioactive macromolecules. *Biopolymers*, *94*, 763–770.

Nava-Parada, P., Forni, G., Knutson, K. L., Pease, L. R., & Celis, E. (2007). Peptide vaccine given with a Toll-like receptor agonist is effective for the treatment and prevention of spontaneous breast tumors. *Cancer Research*, *67*, 1326–1334.

Neoptolemos, J. P. (2014). Predictive cytokine biomarkers for survival in patients with advanced pancreatic cancer randomized to sequential chemoimmunotherapy comprising gemcitabine and capecitabine (GemCap) followed by the telomerase vaccine GV1001 compared to concurrent chemoimmunotherapy in the TeloVac phase III trial. In *2014 ASCO Annual Meeting, Chicago, USA*.

Nguyen, C. T., Hong, S. H., Sin, J. I., Vu, H. V., Jeong, K., Cho, K. O., et al. (2013). Flagellin enhances tumor-specific CD8(+) T cell immune responses through TLR5 stimulation in a therapeutic cancer vaccine model. *Vaccine*, *31*, 3879–3887.

Ohtake, J., Ohkuri, T., Togashi, Y., Kitamura, H., Okuno, K., & Nishimura, T. (2014). Identification of novel helper epitope peptides of Survivin cancer-associated antigen applicable to developing helper/killer-hybrid epitope long peptide cancer vaccine. *Immunology Letters*, *161*, 20–30.

Ott, P. A., Hodi, F. S., & Robert, C. (2013). CTLA-4 and PD-1/PD-L1 blockade: New immunotherapeutic modalities with durable clinical benefit in melanoma patients. *Clinical Cancer Research*, *19*, 5300–5309.

Pal, S. K., Hu, A., & Figlin, R. A. (2013). A new age for vaccine therapy in renal cell carcinoma. *Cancer Journal*, *19*, 365–370.

Panza, F., Solfrizzi, V., Imbimbo, B. P., Tortelli, R., Santamato, A., & Logroscino, G. (2014). Amyloid-based immunotherapy for Alzheimer's disease in the time of prevention trials: The way forward. *Expert Review of Clinical Immunology*, *10*, 405–419.

Park, H. H., Lee, K. Y., Kim, S., Lee, J. W., Choi, N. Y., Lee, E. H., et al. (2014). Novel vaccine peptide GV1001 effectively blocks beta-amyloid toxicity by mimicking the extra-telomeric functions of human telomerase reverse transcriptase. *Neurobiology of Aging*, *35*, 1255–1274.

Parmiani, G., Castelli, C., Dalerba, P., Mortarini, R., Rivoltini, L., Marincola, F. M., et al. (2002). Cancer immunotherapy with peptide-based vaccines: What have we achieved? Where are we going? *Journal of the National Cancer Institute*, *94*, 805–818.

Parry, R. V., Chemnitz, J. M., Frauwirth, K. A., Lanfranco, A. R., Braunstein, I., Kobayashi, S. V., et al. (2005). CTLA-4 and PD-1 receptors inhibit T-cell activation by distinct mechanisms. *Molecular and Cellular Biology*, *25*, 9543–9553.

Patel, D., Couroux, P., Hickey, P., Salapatek, A. M., Laidler, P., Larche, M., et al. (2013). Fel d 1-derived peptide antigen desensitization shows a persistent treatment effect 1 year after the start of dosing: A randomized, placebo-controlled study. *The Journal of Allergy and Clinical Immunology*, *131*, 103-9.e1–103-9.e7.

Patton, R. L., Kalback, W. M., Esh, C. L., Kokjohn, T. A., Van Vickle, G. D., Luehrs, D. C., et al. (2006). Amyloid-beta peptide remnants in AN-1792-immunized Alzheimer's disease patients: A biochemical analysis. *The American Journal of Pathology*, *169*, 1048–1063.

Pellaton, C., Perrin, Y., Boudousquie, C., Barbier, N., Wassenberg, J., Corradin, G., et al. (2013). Novel birch pollen specific immunotherapy formulation based on contiguous overlapping peptides. *Clinical and Translational Allergy*, *3*, 17.

Petrovsky, N., & Aguilar, J. C. (2004). Vaccine adjuvants: Current state and future trends. *Immunology and Cell Biology*, *82*, 488–496.

Pilon, J. L., Rhyan, J. C., Wolfe, L. L., Davis, T. R., McCollum, M. P., O'Rourke, K. I., et al. (2013). Immunization with a synthetic peptide vaccine fails to protect mule deer (Odocoileus hemionus) from chronic wasting disease. *Journal of Wildlife Diseases*, *49*, 694–698.

Purcell, A. W., McCluskey, J., & Rossjohn, J. (2007). More than one reason to rethink the use of peptides in vaccine design. *Nature Reviews. Drug Discovery*, *6*, 11.

Rausch, S., Kruck, S., Stenzl, A., & Bedke, J. (2014). IMA901 for metastatic renal cell carcinoma in the context of new approaches to immunotherapy. *Future Oncology*, *10*, 937–948.

Raval, R. R., Sharabi, A. B., Walker, A. J., Drake, C. G., & Sharma, P. (2014). Tumor immunology and cancer immunotherapy: Summary of the 2013 SITC primer. *Journal for Immunotherapy of Cancer*, *2*, 14.

Rezvani, K., Yong, A. S., Mielke, S., Jafarpour, B., Savani, B. N., Le, R. Q., et al. (2011). Repeated PR1 and WT1 peptide vaccination in Montanide-adjuvant fails to induce sustained high-avidity, epitope-specific CD8+ T cells in myeloid malignancies. *Haematologica*, *96*, 432–440.

Serrano-Pozo, A., Frosch, M. P., Masliah, E., & Hyman, B. T. (2011). Neuropathological alterations in Alzheimer disease. *Cold Spring Harbor Perspectives in Medicine*, *1*, a006189.

Sigurdsson, E. M. (2008). Immunotherapy targeting pathological tau protein in Alzheimer's disease and related tauopathies. *Journal of Alzheimer's Disease*, *15*, 157–168.

Su, Y., Rossi, R., De Groot, A. S., & Scott, D. W. (2013). Regulatory T cell epitopes (Tregitopes) in IgG induce tolerance in vivo and lack immunogenicity per se. *Journal of Leukocyte Biology*, *94*, 377–383.

Takahashi, N., Ohkuri, T., Homma, S., Ohtake, J., Wakita, D., Togashi, Y., et al. (2012). First clinical trial of cancer vaccine therapy with artificially synthesized helper/killer-hybrid epitope long peptide of MAGE-A4 cancer antigen. *Cancer Science*, *103*, 150–153.

Theunis, C., Crespo-Biel, N., Gafner, V., Pihlgren, M., Lopez-Deber, M. P., Reis, P., et al. (2013). Efficacy and safety of a liposome-based vaccine against protein Tau, assessed in tau.P301L mice that model tauopathy. *PLoS One*, *8*, e72301.

Walter, S., Weinschenk, T., Reinhardt, C., & Singh-Jasuja, H. (2013). Single-dose cyclophosphamide synergizes with immune responses to the renal cell cancer vaccine IMA901. *OncoImmunology*, *2*, e22246.

Walter, S., Weinschenk, T., Stenzl, A., Zdrojowy, R., Pluzanska, A., Szczylik, C., et al. (2012). Multipeptide immune response to cancer vaccine IMA901 after single-dose cyclophosphamide associates with longer patient survival. *Nature Medicine*, *18*, 1254–1261.

Worm, M., Patel, D., & Creticos, P. S. (2013). Cat peptide antigen desensitisation for treating cat allergic rhinoconjunctivitis. *Expert Opinion on Investigational Drugs*, *22*, 1347–1357.

CHAPTER TWO

The Use of Liposomes to Shape Epitope Structure and Modulate Immunogenic Responses of Peptide Vaccines Against HIV MPER

Beatriz Apellániz*,†, José L. Nieva*,†,1

*Biophysics Unit (CSIC, UPV/EHU), University of the Basque Country (UPV/EHU), Bilbao, Spain
†Department of Biochemistry and Molecular Biology, University of the Basque Country (UPV/EHU), Bilbao, Spain
[1]Corresponding author: e-mail address: gbpniesj@lg.ehu.es

Contents

Abstract

Peptide vaccines have been shown effective in preventing animal infection in some instances, and various formulations are under evaluation for their potential clinical use in humans. In the case of the Human Immunodeficiency Virus type-1 (HIV-1) infection, viral escape from immune surveillance restricts relevant neutralizing humoral responses to a handful of sites of vulnerability on the envelope glycoprotein. The membrane-proximal external region (MPER) on the gp41 transmembrane subunit has been identified as the only linear B-epitope that embodies an HIV vulnerability site. Thus, focusing humoral responses to MPER by peptide-based immunogens is a pursued goal in HIV vaccine development. The location of this sequence in the vicinity of the membrane interface, its composition (rich in aromatic residues), and the requirement of long-hydrophobic heavy-chain third complementarity-determining region loops for antibody-mediated neutralization suggests that in addition to the specific amino

Advances in Protein Chemistry and Structural Biology, Volume 99
ISSN 1876-1623
http://dx.doi.org/10.1016/bs.apcsb.2015.03.002

acid composition, antigenicity and immunogenicity of MPER can be modulated by membrane lipids. In this chapter, we give an overview of applications of lipid vesicles (liposomes) to the development of MPER-targeting vaccines, both as type-B adjuvants and epitope structure-shaping devices.

1. INTRODUCTION

A successful vaccine candidate should meet several criteria: immunogenicity, specificity, protective activity, and durability. First and foremost, it should be immunogenic, i.e., it should be able to elicit an immune response, both humoral- and cell-mediated, that leads to the blocking and eradication of the disease-causing agent and clearance of the affected cells. Second, this immune response must specifically target (bind with a certain affinity) the region on the pathogen that has been mimicked by the immunogen used as a vaccine, that being a peptide, a protein, or a whole organism, and it must avoid cross-reaction with other self-antigens to prevent autoimmunity. Third, the elicited immune response should be able to prevent the establishment of a disease if the aim is to pursue a prophylactic vaccine, which is usually the case in pathogen-mediated infections, including prions (Goni et al., 2015; Plotkin, 2014). Finally, the vaccine must be able to induce the production of B and T memory cells, both required to ensure reelicitation of the protective immune response should the organism encounter the pathogenic agent even years after immunization (Slifka & Amanna, 2014).

To date, the target pathogenic agents only include bacteria and viruses, and the vaccines employed can be classified in two main categories: (i) whole organisms, killed, or live attenuated and (ii) selected antigens of these organisms, such as purified proteins, capsular polysaccharides, conjugates of both, and viral-like particles, some of them genetically engineered (Plotkin, 2014). In contrast to the vaccines in the first category, those in the second group include only the regions of the pathogens toward which the response is targeted. The benefit of this strategy is twofold. On the one hand, since protective immunity usually depends on a few molecules of the pathogen, the risk of adverse reactions including inflammation, autoimmune response, or reversion to virulence is lowered in these vaccines. Furthermore, many pathogens can be difficult or impossible to culture by conventional methods (Black, Trent, Tirrell, & Olive, 2010; Cantor, 1999; Esparza, 2013; Li, Joshi, Singhania, Ramsey, & Murthy, 2014; Purcell, McCluskey, & Rossjohn, 2007).

On the other hand, apart from excluding these potentially harmful elements that do not add to the protective response, the immune system is prevented from being diverted by these regions and thus, proper focusing on the protective immunity inducing ones is ensured (Correia et al., 2014; Tobin et al., 2008). However, the immunogenicity of these subunit vaccines often decreases concomitantly, and appropriate delivery systems and potent immunostimulants have to be included in the formulation to compensate for this fact (Brunner, Jensen-Jarolim, & Pali-Schöll, 2010; Coffman, Sher, & Seder, 2010; Hafner, Corthesy, & Merkle, 2013).

Following this reductionist approach, the next step would be to dissect the subunit vaccine to determine which regions are responsible for inducing the protective immune response. T and B lymphocytes receptors (TCRs and BCRs, respectively) recognize small oligopeptides, called epitopes, continuous in sequence (linear epitopes) in the first case, and continuous or discontinuous (conformational epitopes) in the latter (Dyson, Lerner, & Wright, 1988; Sesardic, 1993). Steric and chemical complementarity between the combining site surface and the bulky side chains protruding from a conformationally constrained backbone, underlie the specific recognition of a continuous amino acid sequence by TCRs or BCRs. A single protein may contain hundreds of epitopes that might lead to T- and B-cell activation. However, the elicited response might not be protective, i.e., these epitopes are immunogenic but the immune response generated is nonneutralizing. Moreover, some of these epitopes might be used as immunological decoys, evolving in pathogens to evade host immune systems. In HIV, for example, key viral epitopes on the Env spike are protected by the presence of a shield of glycans, adjacent variable immunodominant loops, and conformational masking (Burton, Stanfield, & Wilson, 2005; Kwong et al., 2002; Wei et al., 2003). Thus, vaccination strategies using the whole protein have proven ineffective and strictly limited to the strain represented by the vaccine (Barouch, 2007; Esparza, 2013; Walker & Burton, 2008; Watkins, 2008). Therefore, it is proposed that only the minimal epitopes required to elicit a protective response should be included in a vaccine, an idea that has led to the development of synthetic peptide vaccines (Black et al., 2010; Cantor, 1999; Caoili, 2013; Francis, 2003; Li et al., 2014; Purcell et al., 2007; Sesardic, 1993).

Specifically, a peptide vaccine consists of one or more peptide sequences of more than 15 amino acids long that induce B- and T-cell stimulation when presented by itself or bound to carrier proteins, scaffolds, or supramolecular complexes (e.g., liposomes, see below; Alving, Rao, Steers,

Matyas, & Mayorov, 2012; Francis, 2003; Lee, Huang, Jayathilaka, Lateef, & Gupta, 2010; Li et al., 2014; Moisa & Kolesanova, 2011; Rosendahl Huber, van Beek, de Jonge, Luytjes, & van Baarle, 2014). Moreover, peptides from one or several strains of the same pathogen might be included to ensure proper coverage of pathogen variability. Some peptide-based formulations devised following these strategies have rendered vaccines effective in preventing viral infection in animals (Bittle et al., 1982; Langeveld et al., 1994).

The selected peptides represent immunogenic epitopes of the native protein that may be linear or conformational but, in either case, they have to elicit an immune response that cross-reacts with the target protein in the pathogenic agent and interfere with its biological function (Caoili, 2013). Consequently, not only the sequence but also the structure of the peptide should resemble that of the epitope in the cognate molecular target. In this chapter, we introduce a new paradigm in peptide vaccine development: the structural stabilization of peptide epitopes through contacts with the membrane surface.

2. MPER: A SITE OF VULNERABILITY ON HIV-1 ENV AND A LINEAR EPITOPE

As mentioned above, peptide-based vaccines could be of particular relevance to prevent Human Immunodeficiency Virus type-1 (HIV-1) infection. A preventive HIV-1 vaccine should stimulate specific responses against vulnerable sites, while avoiding production of antibodies against immunodominant/hypervariable regions exposed on the main antigen, the Env glycoprotein (Fig. 1). Current methodological developments to obtain Env-reactive antibodies have led to the fine definition of the conserved "sites of vulnerability" targeted by broadly neutralizing antibodies (bNAbs; for recent reviews, see: Klein, Mouquet, et al., 2013; Kwong & Mascola, 2012; West et al., 2014). Emergence of bNAbs exerting broad and potent neutralizing activity across different HIV-1 strains and primary isolates requires years in contact with the antigen, but their binding is restricted to a handful of sites on the Env complex. Thus, HIV bNAbs can be considered natural products optimized under selective pressure to recognize and bind vulnerable epitopes. These major HIV-1 sites of vulnerability include the receptor-binding site, the glycan-V3 site, and the V1/V2 region on the surface gp120 subunit, and the "membrane-proximal external region" (MPER) on the transmembrane gp41 subunit. More recently, new

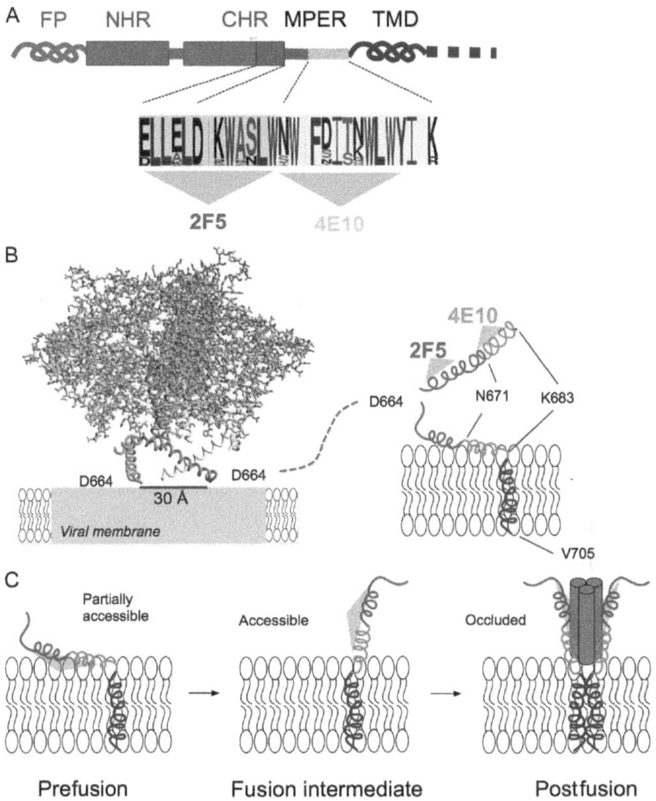

Figure 1 Designation of the HIV-1 linear epitope MPER and its membrane interaction modes. (A) Diagram of the Env gp41 subunit. The highlighted elements include: FP, fusion peptide; NHR and CHR, amino- and carboxy-terminal helical regions, respectively; MPER, membrane-proximal external region; TMD, transmembrane domain. MPER sequence variability within HIV-1 clade B has been displayed below as a WebLogo representation (Crooks, Hon, Chandonia, & Brenner, 2004). Nonpolar amino acids are blue colored (dark gray color in the print version). Positions of the extended 2F5 and 4E10 epitopes are also indicated. (B) The structure of the SOSIP gp140 HIV-1 Env trimer (PDB ID: 4NCO) and its position relative to the viral membrane is shown in the left panel. Gp120 subunits are depicted in stick representation (orange; dark gray color in the print version). Helical regions of gp41 trimers are shown in cyan (dark gray color in the print version), pink (light gray color in the print version), and gray colors. Except for an N-terminal stretch ending at position Asp-664, the juxtaposed MPER is not included in the crystal structure. The model on the right emphasizes MPER folding in contact with the viral membrane. (C) Exposure degrees proposed for the neutralizing face of MPER along the fusion process. Epitopes within a helix adsorbed to the membrane interface would be partially accessible. Maximal exposure would occur upon fusion activation, within structurally poorly defined intermediates. Epitopes get totally occluded within postfusion structures (6-helix bundles).

bNAbs binding to epitopes conformed within the matured, oligomeric Env structure have been established (Blattner et al., 2014; Falkowska et al., 2014; Huang et al., 2014; Scharf et al., 2014).

The MPER comprises an aromatic-rich sequence connecting HIV-1 gp41 ectodomain to the transmembrane domain (TMD) anchor (Fig. 1A). In contrast to the structurally complex, discontinuous epitopes recognized by anti-gp120 bNAbs, or bNAbs targeting the Env quaternary structure, it is hypothesized that MPER embodies a single continuous linear epitope (Cardoso et al., 2007; Huang et al., 2012; Ofek et al., 2004; Pejchal et al., 2009; Zhu et al., 2011). Structural evidences support this assumption. Recent biochemical and electron microscopy studies demonstrate that MPER contributes marginally to stabilize the Env trimeric complex (Khayat et al., 2013; Klasse et al., 2013), thereby suggesting that MPER does not take part in tertiary or quaternary interactions with upstream ectodomain sequences. In addition, the crystal structures of the native Env complex confirm the absence of interprotomer interactions at the level of the MPER D664 residue (Julien et al., 2013; Pancera et al., 2014; Fig. 1B; left). Hence, structural data reinforce the idea that the linear MPER sequence downstream D664 folds independently in the vicinity of the viral membrane (Fig 1B; right).

Systematic analyses of sera collected from chronically infected individuals reveal that some HIV-1 neutralizing activity can be mapped to the gp41 MPER domain, thus supporting the functionality of anti-MPER antibodies during infection (Gray et al., 2009; Li et al., 2009; Sather & Stamatatos, 2010; Shen et al., 2009; Stamatatos, Morris, Burton, & Mascola, 2009; Zhou et al., 2012; Zhu et al., 2011). As an example, analysis of the sera from 78 healthy HIV-1-infected donors exhibiting neutralizing activity demonstrated that 27% contained MPER-specific antibodies (Huang et al., 2012). Interestingly, among the isolated anti-MPER bNAbs (Table 1), those directed against the MPER C-terminal region have been recognized as unusual antibodies displaying near pan-neutralizing activity in standard panels of HIV strains and isolates (Binley et al., 2004; Huang et al., 2012; Kunert, Ruker, & Katinger, 1998; Muster et al., 1993, 1994; Purtscher et al., 1994; Stiegler et al., 2001).

Further supporting their functional activity *in vivo*, anti-MPER bNAbs confer protection to viral infection when passively transferred to primate models (Ferrantelli et al., 2003) and increase the neutralization response upon passive immunization in humans (Stiegler et al., 2002; Trkola et al., 2005). Remarkably, in one comparative study, the anti-gp41 antibodies

Table 1 Broadly Neutralizing HIV Antibodies Binding to MPER Epitopes

Epitope Location	Antibody Designation	PDB ID	Extended Epitope	Neutralization Breadth[a]
N-terminal	M66/M66.6	4NRX/4NRZ	^{656}NEQELLELDKWASL^{669}	24%
N-terminal	2F5	1TJI; 3D0L	^{656}NEQELLELDKWASLW^{670}	55%
C-terminal	Z13/Z13e1	3FN0	^{668}SLWNWFDITN^{677}	N.D.
C-terminal	4E10	2FX7	^{671}NWFDITNWLWYIK^{683}	98%
C-terminal	10E8/7H6	4G6F	^{668}SLWNWFDITNWLWYIK^{683}	98%
C-terminal	CH12	N.A.	^{670}WSWFDITKWLW^{680}	27%

[a]Percentage of neutralized virus isolates. Data comparing M66.6 and 2F5 were taken from Zhu et al. (2011). 4E10 and 10E8 values as reported by (Huang et al., 2012). CH12 antibody value was based on isolates from the clade C (Morris et al., 2011).

were overall less effective than the anti-gp120 antibodies in neutralization assays, but provided the strongest protection *in vivo*, a property that might depend on the specific characteristics of the MPER epitope (van Gils & Sanders, 2014).

Taking these evidences into account, it has been argued that a vaccine with the capacity to induce a preexisting anti-MPER neutralizing response would have the potential to protect against HIV-1 infection (Stamatatos et al., 2009). However, the numerous attempts to induce high titers of MPER-targeting bNAbs through vaccination have failed so far (Montero, van Houten, Wang, & Scott, 2008). Essentially, two competing theories have been proposed to explain the lack of success of MPER-targeting vaccines. Based on the cross-reactivity with lipids displayed by anti-MPER antibodies, some authors argue that B-cells producing such antibodies would be self-reactive and deleted during differentiation (Haynes et al., 2005; Verkoczy, Kelsoe, & Haynes, 2014). In the context of this theory, self-mimicry arises as an additional mechanism for viral escape. However, this concept has been challenged as a general rule. Some results question the existence of a relationship between the reported lipid polyreactivities and the requirement of self-tolerance break for development of cognate anti-MPER antibodies (Alving, 2008; Santiveri, Rico, Jimenez, Pastor, & Perez-Paya, 2003; Santiveri, Santoro, Rico, & Jimenez, 2004; Scherer, Zwick, Teyton, & Burton, 2007). It is also recognized that some infected individuals raise responses that can neutralize HIV-1 via antibodies to the MPER (Huang et al., 2012; Sather & Stamatatos, 2010). Finally, the isolation of the bNAb 10E8, which neutralizes the virus with exceptional breath and potency in the absence of autoreactivity (Huang et al., 2012), argues against the self-tolerance break requirement for anti-MPER antibody elicitation.

An alternative theory postulates that conserved epitopes within MPER can be immunogenic, but we currently do not understand the viral or host factors underlying efficient development of antibodies targeting this region (Stamatatos et al., 2009). A consequence of this limited understanding has been the incapacity to reconstruct the viral structures recognized by anti-MPER bNAbs to exert neutralization. Assumption of our limited knowledge also implies that research efforts to develop anti-MPER vaccines must intensify by incorporating new strategies.

One of the biggest hurdles that apply to peptide vaccine design in general and to MPER in particular is that, even if an antibody response is generated, it might not recognize the intended viral target due to conformation and other structural differences (e.g., accessibility) between synthetic peptides and epitopes in the native Env complex (Caoili, 2013). Moreover, other residues in the Env glycoprotein, not included in the peptide vaccine, might affect the binding of the neutralizing antibodies and the immunogenicity of the epitopes (Kim, Leaman, & Zwick, 2014; Lee et al., 2010; Montero et al., 2012; Van Regenmortel, 2014).

An extra obstacle to be overcome is that HIV bNAbs are generated only after extensive rounds of maturation from germ-line versions of BCRs (somatic hypermutation) (Klein, Diskin, et al., 2013). The epitopes recognized by germ-line versions and the epitopes designed to bind the final bNAbs might not share the same features, and therefore, vaccination with the later might not guide the required maturation process leading to the production of the cognate neutralizing antibody (Klein, Mouquet, et al., 2013; Kulp & Schief, 2013; Van Regenmortel, 2014; West et al., 2014). Thus, somatic hypermutation would pose a challenge for elicitation of MPER antibodies through peptide vaccination (Kersten & Crommelin, 2003; Santiveri, Rico, & Jimenez, 2000; Zhu et al., 2011).

Nonetheless, molecular recognition of the MPER linear epitope is probably less dependent on maturation processes than the rest of conformational HIV epitopes. MPER peptide vaccines can potentially overcome the poor immunogenicity of this region within the Env complex by ensuring its full exposure to the partners of the immune system. Thus, at least theoretically, MPER peptide vaccines could be used to expand specific B-cell clones primed by structurally relevant, but otherwise less immunogenic, subunit vaccines. In this regard, based on the hypothesis that the viral membrane interface plays a crucial role in chaperoning MPER folding, unprecedented liposome-based methodologies for developing MPER-targeting immunogenic formulations have been proposed in the last years (Table 2).

3. STRUCTURAL ORGANIZATION OF MPER

3.1 Hydrophobicity Distribution

Wimley and White (WW) calculated the water-to-membrane interface transfer free energies for each amino acid in the context of unfolded peptide chains (ΔG_{wiu}-s) (White & Wimley, 1999). These authors observed that aromatics (Trp, Tyr, and Phe) are the most hydrophobic-at-interface residues present in proteins. The list of free energy values of partitioning were proposed as the WW "interfacial hydrophobicity scale." When applied to the gp41 MPER-TMD region, the classical Kyte–Doolitle (KD) hydropathy index reveals a positive peak spanning the gp41 TMD

Table 2 Immunization Assays Using Liposome-MPER as Immunogen

Immunogen	Adjuvant	Animal Immunization: Model/ Route/Strategy	Immunogenicity[a]	Neutralization	References
(2) HIV-1 rgp160BaL DNA 12 mer 2F5 epitope (aa 661–675 HXB2) from gp41 subtypes A–D gp41/MN coiled-coil aa 578–591 (3) 2F5 epitope peptides gp41/MN coiled-coil peptide human CCR5 peptide aa 168–182	Monooleate/fatty acid	C57BL/6 mice i.n. (2) 1 DNA prime and 1 (DNA/2F5ep/coiled coil) boost (3) 1 prime and 1 boost (homologous: 3 peptides, twice)	Ag 2F5ep (A–D) (2) 3500–5000 (3) 1600–3700	ID80: 1/20–1/80 Negative	Devito et al. (2004)
CHR-MPER[b]-TM aa 629–706 (crosslinked trimer, gp41ctm) in proteoliposomes	Liposomes[c] L-α-PC:L-α-PS:Chol (1:2:1)[c]	BALB/c mice i.p. 1 prime and 1 boost	Ag gp41ctm Control w/o liposomes: 250–640 Liposomes + gp41ctm: 1280–2560	(−) in serum	Lenz et al. (2005)
Same as in Karasavvas et al. (2008)	Tetanus Toxin Liposomes–MLPA DMPC:Chol:PIP (1:1.5:1)	BALB/c mice 1 i.p. prime and 1 i.v. boost	N.D. Specificity: IgM mAb WR309, recognition of mper48 and liposomes	N.D.	Beck et al. (2008)
TetH-MPER48-kk QYIKANSKFIGITEL-GGGG-^{636}MPER^{683}	Tetanus Toxin Liposomes–MLPA* DMPC:Chol:DMPG: GalCer (9:7.5:1:0.13)	BALB/c mice 1 i.p. prime and 1 i.v. boost	Ag MPER48[d] 400–800 Specificity: negative (no recognition of ^{661}MPER^{683} by purified IgM monoclonals)	N.D.	Karasavvas et al. (2008)

MPER N674F aa 662–684	Liposomes–MLPA DMPC:Chol:PIP (1:1.5:1)	BALB/c mice i.p. 1 prime and 3 boosts	Ag MPER immunogen 2×10^4 Specificity: 2 purified monoclonal IgM: PIP/PIP2 and WR316 core: SLWNWF WR320 core: LELDKWASL	IC50 WR320 IgM MAb 20–100 μg/mL	Matyas et al. (2009)
N-MPER-GGK-X aa 656–671 X = -CHEMS (1), -palm (2), -PC (3), -DPG (4), -CL (5), -Chol (6), -SM (7), -PE (8), unconjugated (9) C-MPER-AAK-X aa 671–683 X = -CHEMS (10), -palm (11), -PC (12), -DPG (13), unconj. (14) Z-(NC-MPER-BKBK)-X aa 656–683 Z,X = unconj., -CHEMS (15), unconj., -DPG (16), unconj, unconj (17), -CHEMS, unconj (18), -CHEMS, -CHEMS (19)	Liposomes–MLPA DMPC:DMPG:Chol:MPLA (15:2:3:0.3)	BALB/c mice s.c. 1 prime and 1 boost	Ag N-MPER (2F5 ep.) (1) 6.4×10^4 but (−) w/o MLPA (2) and (3) negative (4)–(8) $7–40 \times 10^3$ (9) 400 (2 out of 5 mice) (15) 2.5×10^5 (17) 1.5×10^5 (18) 4000 (19) 7500 Ag C-MPER (4E10/10E8 ep.) (11)–(14) and (17) negative (10) and (18) 600 (15) 900 (19) 300	N.D.	Watson and Szoka (2009)
JRFLgp140CF (^{675}D, ^{678}K) (1) and (6) MPER aa 652–671 (2) and (7) JRFLgp140CF&MPER aa 656–671 (3) and (4) JRFLgp140CF&MPER aa 656–683 (5) and (8) (^{675}N, ^{678}N)	gp140: Emulsigen/oCpG MPER: Liposomes–MLPA and Emulsignen/oCpG POPC:POPE:DMPA:Chol (45:25:20:1.33)	Guinea pigs (2)–(4) i.m./i.n./s.l. (1) and (5) i.m. (1) and (2) 1 prime and 2 boost (3) 3 MPER–1 gp140–3 MPER (4) 1 gp140–3 MPER (5) 4 MPER–1 gp140–3 MPER–1 gp140–2 MPER	Specific IgG in serum by SPR Ag MPER aa 562–671; 1/50 dilution: (1) 30 RU; specificity: ^{666}W (1/3 animals) (2) 600–1000 RU; specificity: ^{664}DK (1/3) (3) 1000–1500 RU: specificity: D or K (4) 500–2000 RU: specificity: DK (5) 100–200 RU: specificity: DKW (1/3) recipr. endpoint titer by ELISA: 10^5	N.D. in guinea pigs	Dennison et al. (2011)

Continued

Table 2 Immunization Assays Using Liposome-MPER as Immunogen—cont'd

Immunogen	Adjuvant	Animal Immunization: Model/ Route/Strategy	Immunogenicity	Neutralization	References
	gp140: AS01 MPER: Liposomes-MLPA/ R848 and Emulsigen/oCpG	Rhesus macaques/i.m. (6) 5 gp140 (7) 5 MPER (8) 5 gp140–3 MPER	(6) Recipr. endpoint titer by ELISA: $<10^2$ (7) 900–1400 RU: specificity D (8) 500–2000 RU: specificity DKW (3/4) recipr. endpoint titer by ELISA: 10^3–10^4	(−) in serum	
N-MPER-GGKGG-H_6 aa 656–671 + X X = DOGS-mono-NTA (1), DOD-tris-NTA (2), free peptide (3) N-MPER-GGK-CHEMS aa 656–671 (4)	Same as in Watson and Szoka (2009)	Same as in Watson and Szoka (2009)	Ag N-MPER (2F5 ep.) (1) and (2) 550–760 (3) negative (4) 4.4×10^4	N.D.	Watson et al. (2011)
FPp-kkk aa 512–534 and MPERp aa 656–683	MDP (prime) Liposomes (prime and boost) POPC:Chol:PA (2:1.5:0.2)	Rabbit / i.d. 1 prime and 4 boosts	Ag MPERp (2F5 ep.) 10^3–10^4	IC50; purified polyclonal anti-^{656}MPER^{671} Ab 22.5 μg/mL	Huarte et al. (2012)
LP-GCN4-ollas-G4S-MPER*-TM-CT_{VSVG} (1) *MPER42-TM: aa 641–710 (2) *MPER24-TM: aa 660–710 (3) *MPER18-TM: aa 666–710 (4) group 1 w/o GCN4 (5) group 2 w/o GCN4	VLP	BALB/c mice 3 VLP boost (s.c.) with or w/o 2/3 DNA prime (i.m)	Specific IgG in serum by ELISA: Ag ^{656}MPER^{683}; 1/100 serum dilution : Low: groups (4) and (5) Medium: group (3) and w/o DNA prime High: group 1 — Specificity: 2F5 epitope	(−) in serum	Kamdem Toukam et al. (2012)
(1) MPER aa 662–683 (HXB2, 089CON, PB7) (2) N-palm-MPER (3) N-palm-7 spacer-MPER (4) MPER-C-palm (5) MPER-kk-C-palm (6) MPER-TM	Liposomes–MLPA (a) DOPC:DOPG (72:18) (b) DSPC:DSPG (c) DOPC:DOPG:Chol With (e) or w/o 10% PEG (f) LACK1 (g) or PADRE (h)	BALB/c mice I.d. 1 prime and 2 boosts	(1) < (2) both with a, f, g (2) < (3) (") (e) = (f) both with 3, a, g (b) < (a) < (c) (") (h) < (g) both with 3, a, f (4) < (5) } Cterm domain (6) Specificity: Nterm domain	(−) with 100 μg/mL polyclonal total IgG	Kim et al. (2013)

CHR-MPER-TM aa 629–706 (trimer) Proteoliposomes	Liposomes POPC:POPE:POPS:CL/SM: Chol (1:2:1:2:5)	Llamas i.m. 1 prime and 5 boosts	Positive in serum, not shown Isolation of 2H10 VHH single chain Ab: Specificity: ^{657}E, ^{658}Q, ^{661}L, ^{662}E, and ^{665}K	(−) in serum (+) Bivalent 2H10: 14 clade B, 1 A; 8 tier 1, 7 tier 2. IC50 from equal to 2F5 MAb to 100 fold less Dependent on CDR-H3	Lutje Hulsik et al. (2013)
MPER-GG/AA-K-CHEMS with side chain anionic modifications: MPER aa 656–677 (1) MPER aa 656–677 ^{676}T-PO_3 (2) MPER aa 656–671 ^{668}S-PO_3 (3) MPER aa 671–683 ^{680}Y-NO_2 (4) MPER aa 668–683 ^{668}S-PO_3 (5)	Same as in Watson and Szoka (2009) + CFA/IFA	Rabbits s.c. 1 prime + CFA and 5 boosts + IFA	Ag: each peptide immunogen (3)–(5): 100–200 (1): 3600 (2): 6.1×10^4 Specificity (1) and (2): MPER aa 656–677	(−) in serum	Venditto et al. (2013)
JRFLgp140CF and MPER aa 656–683	gp140: Emulsigen/oCpG MPER: Liposomes-MLPA/ R848 and Emulsigen/oCpG POPC:POPE:DMPA:Chol (45:25:20:1.33)	Opossums i.m./i.n./s.l. 2 gp140 prime/3 MPER boost C57BL/6 Mice: i.p.	In opossums: Ag ^{659}MPER^{678} (2F5 ep.) (+) 3.41×10^5; specificity ^{664}D Ag ^{668}MPER^{683} (4E10 ep.) (−) In mice: Ag: ^{656}MPER^{683} : $<10^2$	(−) in serum N.D.	Yang et al. (2013)
JRFLgp140CF MPER-GTH1: ^{656}MPER^{683}—YKRWIILGLNKIVRMYS	Same as in guinea pigs in Dennison et al. (2011)	C57BL/6 deficient 2F5 knock-in mice/i.p. 1 gp140 prime and 5 MPER boost	Ag ^{659}MPER^{678} (2F5 ep.); ID50: 1/1600 (+++) Liposome/Emulsigen (−) Liposomes/MLPA and Emulsigen/ oCpG (−) MPER and Emulsigen/oCpG (+/−) MPER and MLPA and Emulsigen/oCpG (+/−) MPER and alum (+)	ID50: 1/1800 IC50: 0.5 μg/mL	Verkoczy et al. (2013)

Continued

Table 2 Immunization Assays Using Liposome-MPER as Immunogen—cont'd

Immunogen	Adjuvant	Animal Immunization: Model/ Route/Strategy	Immunogenicity	Neutralization	References
X-CHR-MPER-TM aa 629–706 X: none (1); GCN4 (2) and H3 (3)	Carbopol-791P (boost) VLPs	Rabbits/i.m. 2 DNA prime and 2 VLP boost	Ag MPER: ^{654}EQNEKDLLALDS WNNLWNWFDITKWLWYIK^{683} (1),(2) and (3): 0–200	ID50 up to 1/100 in (1) and (2)	Benen et al. (2014)
gp41int-cys GCN4-^{584}Gp41^{684}-cys proteoliposomes gp140CA018	Liposomes PS:PC:PE:SM:PSE:Chol (1:1:1:2:0.5:4.5) Carbopol-971P MF59	Guinea pigs/i.m. 1 prime and 3 boosts (1) gp140CA018 (2) gp41int-cys (3) 1 gp140CA018 and 2 gp41int-cys	Ag kkk-^{656}MPER^{682}-RKKK (1) (−) (2) 10^3–10^4 (3) N.D.	1 clade A, 4 B, 3 C; 6 tier 1, 2 tier 2 (1) ID50: all 10^2–10^6 (2) and (3) ID50 except SF162.LS 50–3000	Lai et al. (2014)
MPER Standard (S): LL(ELDKWA)$_1$SLWNWF Modified (M): LL(ELDKWA)$_3$SLWNWF	Liposomes (SUV) Defensin peptides HBD 2 (a) or 3 (b)	Mice BALB/c and C57BL/6 i.n. 1 prime and 2 boost	Ag MPER, Endpoint titer × 10^{-3} IgG IgA S 6.4 1.6 Sa 25.6 6.4 Sb 51.2 6.4 M 51.2 25.6 Ma 102.4 51.2 Mb 204.8 51.2 Empty Liposomes 0.2 0.1	1/8 dilution (% neutr): 12 65 70 60 75 85	Mohan, Verma, and Rao (2014)
MPERp aa 656–683	Liposomes–MLPA (1) POPC:Chol:PA (2) POPG	Same as in Huarte et al. (2012)	Same as in Huarte et al. (2012)	IC50; purified polyclonal anti-^{656}MPER^{671} Ab (1) negative (2) 100 μg/mL	Serrano et al. (2014)
Same as in Verkoczy et al. (2013)	Same as in Verkoczy et al. (2013)	Mice/i.m. or i.p B10.D2.S (H-2^s) C57BL/6, BALB.B, B10 (H-2^b) and B10.D2.BR (H-2^k) BALB/c mice (H-2^d) 1 prime and 5 boosts	Ag ^{659}MPER^{678} (2F5 epitope) 0 10^2–10^3 >10^7; specificity ^{665}KW	(−) in serum	Zhang et al. (2014)

N-palm-MPER aa 662–683 *MPER:lipid 1:200 (mol:mol)	Liposomes: 65 nm (a), 150 nm (b), 200 nm (c) (1) DOPC:DOPG (4:1) (2) DMPC:DOPC:DOPG (2:2:1) (3) LACK1 (4) HIV-30 (5) Lipid-PEG-cpG (6) MLPA (7) Cyclic-di-GMP	BALB/c mice s.c. 1 prime and 2 boosts	Ag MPER (a) 700 in (1,3,6) (b) 10^4–10^5 (") (c) 5x10^3 (") (1b) <10^2 in (4) (1b) 10^4–10^5 in (4,6) (1b) 10^4–10^5 in (4,5) (1b) 10^4–10^6 in (4,7) (2b) 10^5–10^6 in (4,6) *Optimal surface density: (MPER:lipid) (1:500 mol:mol) in (1,3,6): 10^4	(−) in serum	Hanson et al. (2015)

[a]Reciprocal endpoint titers of specific IgG in serum by ELISA, unless otherwise noted.
[b]HXB2 numbering, unless otherwise noted.
[c]Large unilamellar vesicles, unless otherwise noted.
[d]Data from Beck et al. (2008).
Immunization routes are abbreviated as follows: intranasal, i.n.; intraperitoneal, i.p.; subcutaneous, s.c.; intramuscular, i.m.; intravenous, i.v.; intradermal i.d.; s.l., sublingual.
Lipids are abbreviated as follows: DM, dimyristoyl; DO, dioleyl; PO, palmitoyloleyl; DS, distearoyl; lPC, phosphatidylcholine; PE, phosphatidylethanolamine; PS, phosphatidylserine; PG, phosphatidylglycerol; Chol, cholesterol; PIP, phosphatidylinositol-4-phosphate; CL, cardiolipin; SM, sphingomyelin; GalCer, galactosylceramide; PSE, dipalmitoylphosphothioethanol; DPG, dipalmitoylglycerol; MLPA, monophosphoryl lipid A; CHEMS, cholesterol hemisuccinate; palm, palmitic acid.
Other abbreviations: Ab, antibody; Ag, antigen; ep., epitope; B, aminoisobutyric acid; CFA/IFA, Complete/Incomplete Freund's Adjuvant; MDP, muramyldipeptide; SUV, small unilamellar vesicle. N.D. no data.

(Lorizate, Huarte, Saez-Cirion, & Nieva, 2008). By comparison, WW detects a positive maximum shifted 15–20 residues toward the N-terminus. The combined WW–KD analysis thus suggests that the nonpolar region (approximately 40–50 residues long) at the C-terminus of the gp41 ectodomain is segmented into two domains: one, corresponding to MPER, positioned somewhat farther from the protein C-terminus (residues ^{664}DKWASLWNWFNITNWLWYIK^{683} according to HXB2 numbering), which would remain stably inserted within the membrane interface, and the other, located closer to the C-terminus, that constitutes the TMD anchor.

Further development of the WW analysis included the calculation of the hydrophobic-at-interface moment (Saez-Cirion et al., 2003). Hydrophobic moments measure the periodicity of residue distribution along the secondary structure elements. Preferential orientation of the hydrophobic residues toward one face of the structure element has been proposed to favor hydrophobic interactions between protein structural components and between membranes and protein sequences. Computation of the hydrophobic moment using the membrane interface-to-water transfer free energies for each amino acid suggests that MPER can be elongated at the N-terminus to include a stretch whose face shows strong affinity for membranes, if folded into an α-helix (residues ^{659}ELLELDKWASLWN^{671}). Underlining the conservation of this structural motif, amphipathic-at-interface helices were detected in different HIV-1 isolates and in other lentiviruses (Saez-Cirion et al., 2003).

In summary, analyses of the hydrophobicity distribution predict the existence of a gp41 region anchored to the viral membrane through an MPER interfacial sequence, which is followed by the fully hydrophobic TMD helix (Fig. 1B; right). Moreover, the MPER region would be segmented into two subdomains: an amphipathic-at-interface helix at the N-terminus and a fully hydrophobic-at-interface stretch at the C-terminus, which approximately overlap the extended 2F5 and 4E10 epitopes, respectively (Table 1).

3.2 MPER Peptide Structure

Further and stronger evidence for the segmented organization of MPER comes from the structural characterization of synthetic peptides by NMR. Peptides devised to comprise the amphipathic N-terminal MPER (Env residues ^{659}ELLELDKWASLWN^{671}) showed a tendency to fold in water solution (Barbato et al., 2003; Biron et al., 2002). However, while some structures disclosed a highly populated 3_{10}-helix conformation (Biron et al., 2002), others comprised an ensemble of local conformational propensities (Barbato et al., 2003).

In contrast, peptides based on the more hydrophobic C-terminal sequence revealed kinked helices in membrane mimics (Fig. 2). A peptide containing the 5 Trp-s present in MPER (Env residues ^{665}KWASLWNWFNITNWLWYIK^{683}) disclosed a helix in the membrane-mimetic dodecylphosphocholine (DPC), better defined at the C-terminal end (Schibli, Montelaro, & Vogel, 2001). Importantly, DPC could efficiently solubilize this peptide, but the structure could not be solved in trifluoroethanol due to poor solubility in this medium.

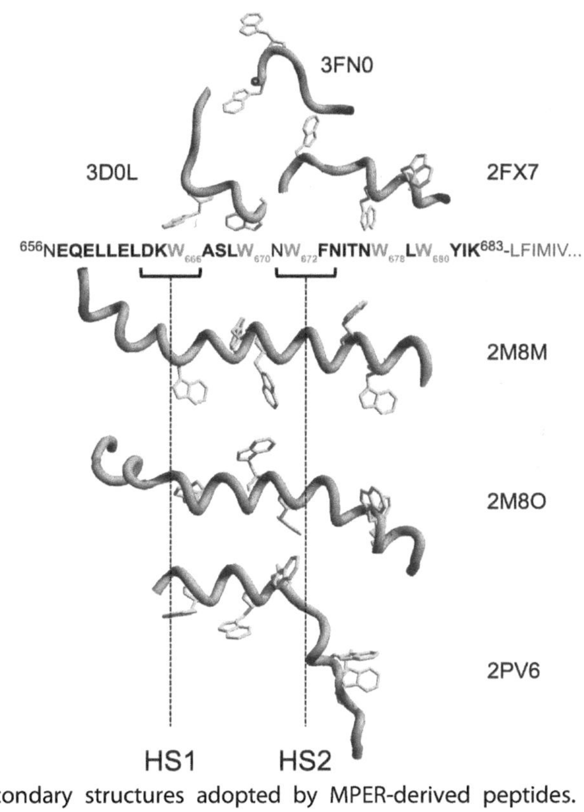

Figure 2 Secondary structures adopted by MPER-derived peptides. PDB accession numbers indicated in the panel designate structures in complex with Fabs 2F5, Z13e1, or 4E10 (3D0L, 3FN0, and 2FX7, respectively), or in contact with DPC micelles (2M8M, 2M8O, and 2PV6). Lateral side chains of Trp residues are depicted in green (light gray in the print version) to align the structures with the MPER amino acid sequence. Proposed positions for hinge segments HS1 and HS2 are also indicated.

This phenomenon was interpreted as evidence that the peptide had a preference for membrane-like environments. A slightly longer peptide (residues ^{662}ELDKWASLWNWFNITNWLWYIK^{683}) adopted a kinked structure in DPC, comprising two segments connected through the short nonhelical ^{673}FN^{674} hinge (Fig. 2, 2PV6; Sun et al., 2008). While the first segment (residues ^{664}DKWASLWNW^{672}) adopted a defined two-turn α-helix conformation, the C-terminal side (residues ^{675}ITNWLWYIK^{683}) was less defined and included a 3_{10}-helical stretch.

More recently, high-resolution structural data pertaining the complete MPER sequence ^{656}NEQELLELDKWASLWNWFNITNWLWYIK^{683} have been reported (Reardon et al., 2014; Serrano et al., 2014). In one study, the NMR structures adopted by an MPER-based peptide (MPERp) in two different membrane-mimetic environments, DPC micelles and 25% HFIP (hexafluoro-2-propanol) were compared (Serrano et al., 2014). MPERp and closely related peptides have been used as constituents of peptide vaccines since they combine within single constructs the full epitopes recognized by M66.6, 2F5, Z13e1, and 4E10/10E8 bNAbs (Alving et al., 2012; Dennison et al., 2011; Kim et al., 2013 Maeso et al., 2011; Serrano et al., 2014; Venditto, Watson, Motion, Montefiori, & Szoka, 2013; Table 1). In HFIP medium, MPERp adopted a continuous helix bent at position ^{664}DKW^{666} (Fig. 2, 2M8M). The sequence also displayed in DPC a well-defined helical structure, but it was kinked at positions ^{661}LEL^{663} and ^{673}FNI^{675} (Fig. 2, 2M8O). A comparable structure was calculated in DPC for a fusion construct expressed in bacteria (Reardon et al., 2014).

In addition, the native-like MPER structures recognized by bNAbs can be inferred from the crystal structures of the antigen-binding fragments (Fabs) complexed with peptide epitopes. The Fab2F5 complexed with peptide revealed a well-defined type I β-turn conformation for ^{664}DKW^{666} core-epitope sequence buried into the paratope (Fig. 2, 3D0L; Julien, Bryson, Nieva, & Pai, 2008). In this structure, the preceding amino-terminal stretch (residues ^{661}LEL^{663}) was fixed into an extended conformation, while the C-terminal side was helical (residues ^{667}ASLW^{670}). By comparison, the sequence recognized by bNAb 4E10 (residues ^{672}WFNITNWLW^{680}) remained overall helical in contact with the Fab (Fig. 2, 2FX7; Cardoso et al., 2007), a pattern also observed for the more recently isolated bNAb 10E8 (Huang et al., 2012). Finally, the FabZ13e1 in complex with the peptide epitope revealed a distinct conformation for the sequence ^{670}WNWFDITN^{677} connecting N- and C-terminal regions of MPER, consisting of two helical turns that are perpendicular to one another (Fig. 2, 3FN0; Pejchal et al., 2009).

These observations provide a comprehensive model for the MPER region as an articulated helical domain, which is distinctly jointed by hinge segments at two positions (Fig. 2). First position (HS1), primarily involves ^{664}DKW^{666}, the critical residues for 2F5 and M66.6 bNAbs binding, but might also include upstream ^{656}NEQELLEL^{663} residues found to adopt α-helix, 3_{10}-helix and fully extended conformations (Barbato et al., 2003; Biron et al., 2002; Julien et al., 2008, 2013; Ofek et al., 2004, 2014; Pancera et al., 2014; Serrano et al., 2014). The second position (HS2, residues ^{671}NWFN/D^{674}) has been defined as a hinge segment with tandem joints in a recent work (Sun et al., 2014). These residues are critical for binding in 4E10, Z13e1, 10E8, and CAP206-CH12 Abs (Cardoso et al., 2007; Huang et al., 2012; Morris et al., 2011; Pejchal et al., 2009; see below). The HS2 segment exhibits conformations compatible with the preservation of an α-helix in the Fab10E8–peptide complex (Huang et al., 2012) and in the MPERp peptide in the presence of HFIP (Fig. 2, 2M8M; Serrano et al., 2014). The α-helix appears kinked by 30° in this position in the MPERp peptide solved in the presence of DPC (Fig. 2, 2M8O; Serrano et al., 2014). A short 3_{10}-helix kink was detected in a Fab4E10–peptide complex (Fig. 2, 2FX7; Cardoso et al., 2007) and the first ever reported NMR structure of MPER (Schibli et al., 2001). More extended hinges lacking canonical conformations were observed for the Z13e1 peptide epitope (Fig. 2, 3FN0; Pejchal et al., 2009) and several peptides solved by NMR in the presence of DPC (Fig. 2, 2PV6; Sun et al., 2008; 2M7W; Reardon et al., 2014). These different conformations support the metastability proposed for the HS2 (Sun et al., 2014), which probably comprises the boundary for two distinct types of MPER epitopes: those located in the N-terminal region (2F5/M66) and those in the C-terminal region (Z13/4E10/10E8; Table 1).

4. LIPID VESICLES: ADJUVANTS AND DEVICES TO SHAPE MPER STRUCTURE

4.1 Liposomes as Adjuvants

As previously stated, the use of small peptides for immunization may encounter the problem of low immunogenicity. Therefore, adjuvants are required to boost peptide immunogenicity and elicit strong and durable responses (Brunner et al., 2010; Coffman et al., 2010). Vesicles made of

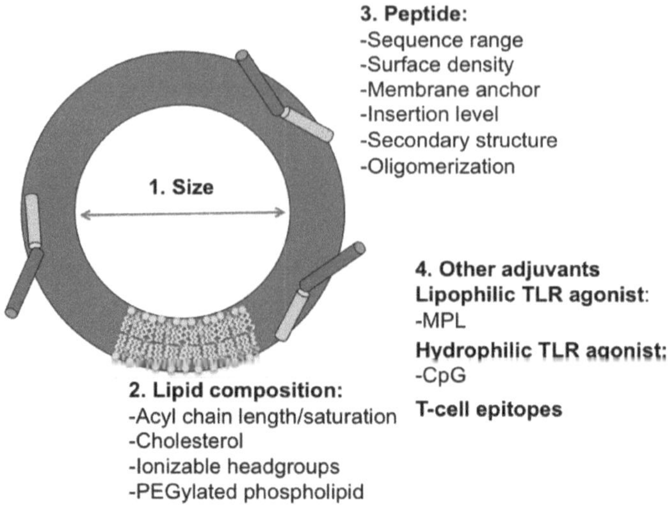

Figure 3 MPER–liposome vaccine components.

phospholipid, or liposomes comprise models that emulate the lipid bilayer architecture of the cell membranes (Fig. 3). As liposome constituents derive from natural membranes, liposomes are safe and well tolerated, i.e., they have a low toxicity, low intrinsic immunogenicity, and are biodegradable. Liposomes can be manufactured to expose on their surface membrane-anchored antigens that resemble repetitive pathogen structures, and/or to encapsulate soluble antigens in the aqueous compartment (for comprehensive reviews, see: Alving et al., 2012; Kersten & Crommelin, 2003; Watson, Endsley, & Huang, 2012). Thus, in principle, liposomes compose type-B adjuvants that enhance antigen presentation by building a depot at the injection site and interacting with antigen-presenting cells in an unspecific manner (Alving et al., 2012; Brunner et al., 2010). However, liposomes can also be subjected to manipulation in order to increase their intrinsic adjuvancy and compose multifunctional vaccines (Alving & Rao, 2008; Kersten & Crommelin, 2003; Watson et al., 2012).

Liposomes are spontaneously generated when diluting cylindrical-shaped amphipathic lipid molecules in aqueous solutions, and depending on the treatment, liposomes can be prepared to have one (unilamellar vesicles) or several bilayers or lamellae (multilamellar vesicles, MLV). The physical properties of the lipid bilayer can determine antigen incorporation and immunogenicity (i.e., degree and pattern of exposure at the surface, encapsulation yield, persistence at the injection site, uptake and trafficking, or isotype switching; Watson et al., 2012). These properties depend on several factors including the vesicle size and lipid composition. As a general rule, unilamellar vesicles with diameters $\geq$100 nm bear bilayer curvatures equivalent to that of the plasma membrane, and are therefore considered as suitable model systems for most studies. These vesicles are generally prepared by mechanical extrusion of MLV suspensions through polycarbonate porous filters of the desired size (Hope, Bally, Webb, & Cullis, 1985).

On the other hand, the elastic properties of the bilayer and the surface charge can be influenced by the chemical nature of the constituent lipids (Kersten & Crommelin, 2003; Watson et al., 2012). Usually, phosphatidylcholine (PC), a zwitterionic phospholipid, is used as the basic phospholipid for liposome preparation. Membrane rigidity can be incremented by choosing phospholipids with high phase transition temperatures (i.e., bearing a high level of acyl chain saturation) or by the inclusion of cholesterol (Chol). In general, less fluid membranes give higher antibody response by the associated antigen. Cationic or anionic liposomal formulations also have superior immunogenic performance when compared to neutral ones. The surface charge can be modified by the incorporation of positively charged stearylamine or negatively charged lipids such as dicetylphosphate, phosphatidylglycerol (PG), or phosphatidylserine (PS). Thermal stability and pH sensitivity can be modulated by including the polymorphic lipid phosphatidylethanolamine and/or cholesteryl hemisuccinate (Hafez & Cullis, 2000), which endow vesicles with the capacity to fuse with membranes of the endocytic compartment and to elicit CTL responses by associated antigen.

Besides the biophysical and biochemical properties of the carrier vesicle, the immunogenicity of liposome-associated antigens depends on the ratio of lipid to antigen and the liposome-antigen attachment mode (Fig. 3, Table 2). Moreover, to have additional effector mechanisms, liposomes can be coadjuvated by incorporating lipophilic and hydrophilic immunostimulatory compounds in the lipid bilayer and the internal aqueous volume, respectively. Different TLR agonists, such as the membrane-associated monophosphoryl lipid A (MPLA) or aqueous markers as unmethylated CpG dinucleotides, have been included into liposomal formulations (Alving & Rao, 2008; Vollmer & Krieg, 2009). Finally, T-cell helper epitopes may be conjugated with surface antigens or encapsulated in the aqueous compartment of the vesicles. Diverse approaches to MPER-liposome vaccine design will be described in the last section of this chapter (see also Table 2).

4.2 The Lipid Bilayer as an MPER Scaffold

Epitope peptides can be synthesized to include nonnatural amino acid modifications, N- and C-terminal caps, other macromolecules like lipids, carbohydrates, carrier proteins and scaffolds, phosphate groups, cyclization moieties or they can be multimerized to improve immunogenicity and stability (Cantor, 1999; Francis, 2003; Moisa & Kolesanova, 2011; Purcell et al., 2007). B-cell epitopes and/or cytolytic T lymphocyte epitopes bearing helper T-cell epitopes might also be added to the main backbone (Li et al., 2014; Testa & Philip, 2012). When available, structural information of both the target protein and the neutralizing antibodies that it elicits *in vivo*, can aid in choosing the epitopes to be mimicked by the vaccine (mimotopes), what is known as structure-based vaccine design or structural vaccinology (Burton et al., 2012; Kulp & Schief, 2013; McLellan et al., 2013) and has been successfully applied recently (McLellan et al., 2013).

In search for structure-based immunogens targeting MPER, initial strategies focused on conformational stabilization of the core-epitopes recognized by 2F5 and 4E10 antibodies. In early reports, the conformation of the 2F5 core-epitope ELDKWAS sequence was constrained either through cyclation of synthetic peptides (McGaughey et al., 2003; Tian et al., 2002), or by grafting it into recombinant protein loops (Coeffier et al., 2000; Ho et al., 2005; Muster et al., 1994). These approaches yielded immunogens that bound with higher affinity to 2F5 monoclonal antibody (MAb), but that were unable to trigger neutralizing responses in animal models. Similarly, to limit the conformational flexibility of the 4E10/Z13e1 peptide epitopes and enhance antibody binding, helix-promoting 2-aminoisobutyric acid residues, and helix-inducing tethers were incorporated into the sequence and shown to improve antibody affinity (Brunel et al., 2006; Ingale, Gach, Zwick, & Dawson, 2010). More recently, the full MPER sequence was subjected to staple scanning in order to stabilize N-terminal (^{662}ELDKWASLWN^{671}) or C-terminal (^{675}ITNWLWYIK^{683}) helices with the orientation found in contact with 4E10/10E8 antibodies (Bird et al., 2014). Peptides with 678–681 residues stapled exhibited enhanced 4E10-binding affinity. When solved in complex with Fab4E10, the crystal structure revealed the presence of a bound phosphate group close to the CDR-H1 loop, thereby suggesting a role of phosphate engagement as a 4E10-binding determinant at the membrane (Bird et al., 2014). A modified stapled peptide containing a tethered phosphate moiety at position 683 exhibited improved binding to 4E10 and was postulated to embody a potential stand-alone peptide vaccine.

Following rational scaffolding approaches, the 2F5, Z13e1, and 4E10 epitope sequences have also been transplanted into proteins exposing peptide chains that assume the conformation of these epitope sequences in complex with the respective antibodies (Correia et al., 2010; Guenaga et al., 2011; Ofek, Guenaga, et al., 2010; Stanfield et al., 2011). As predicted, the epitope-scaffolds (ESs) in complex with the antibodies disclosed the standard epitope conformations for the grafted sequences. Nonetheless, the recovered antibodies after ES immunization did not neutralize the virus. In one of those approaches, the authors performed crystallographic analyses of the 2F5 peptide epitope in complex with MAbs elicited by the ESs (Ofek, Guenaga, et al., 2010). Even though peptide epitopes in complex with parental Fab2F5- and ES-elicited monoclonals adopted comparable conformations and oriented side chains in remarkably similar ways, the ES-elicited antibody lacked a hydrophobic, long CDR-H3 loop, which is essential for 2F5 neutralizing activity (Burton, 2010).

In spite of the different mechanisms that have been proposed to explain the phenomenon, it is commonly accepted that hydrophobic interactions between the CDR-H3 loops and the viral membrane are required for accomplishing the neutralizing activity of anti-MPER 2F5 and 4E10 bNMAbs (Dennison et al., 2011; Julien et al., 2010; Ofek, McKee, et al., 2010; Scherer, Leaman, Zwick, McMichael, & Burton, 2010; Xu et al., 2010). The distribution of polar and nonpolar (aromatic and aliphatic) residues supports the existence of a low-energy MPER structure inserted into the viral membrane external interface (Saez-Cirion et al., 2003; Fig. 1B; right). Recent structural studies of the trimeric SOSIP constructs are consistent with the immersion of native MPER domains into the viral envelope disconnected from each other (Julien et al., 2013; Khayat et al., 2013; Klasse et al., 2013; Lyumkis et al., 2013; Pancera et al., 2014; Fig. 1B; left). This implies that orientation and accessibility of this vulnerable Env site may primarily depend on interactions of MPER residues with the membrane interface plus motional restrictions imposed by the TMD anchor (Montero et al., 2012). In the current model, the MPER helix is predicted to associate with the membrane interface with its axis mainly parallel to the membrane plane. Then, it would bend about 90° at Lys-683 to allow insertion of the TMD in a perpendicular orientation (Fig. 1B; right).

Anti-MPER 2F5, Z13e, and 4E10 antibodies can effectively bind to epitopes buried in the membrane interface (Lorizate, Cruz, et al., 2006; Song et al., 2009). Membrane-inserted MPER, structurally constrained through partitioning-folding coupling, might theoretically encompass a relevant anti-MPER immunogen, i.e., with the potential to generate antibodies with long CDR-H3 loops (Dennison et al., 2009; Huarte et al., 2008). Thus, the membrane could function as a natural scaffold to shape the MPER epitope structure required to generate neutralizing antibodies qualitatively similar to 2F5 and 4E10 (Alving et al., 2012). However, not all membrane-bound states of MPER qualify as 2F5 and/or 4E10 epitope recreations. NMR measurements of MPERps in membrane mimics (DPC micelles) (Schibli et al., 2001; Sun et al., 2008) display lipid-buried structures that differ in conformation and/or orientation from the antibody-bound structures (Cardoso et al., 2005; Julien et al., 2008; Ofek et al., 2004; Fig. 2). Moreover, it can be inferred that MPER structures deeply inserted into the membrane would not engage with BCRs and therefore be devoid of the capacity for generating 2F5 or 4E10-like immune responses (Serrano et al., 2014).

Thus, to explain the high affinity and neutralizing potency displayed by anti-MPER bNAbs, the neutralization competent structures recognized by these antibodies must exist exposed on virion surfaces. In this context, membrane association is required but not sufficient for recreating the MPER structure eliciting neutralizing responses. A native-like architecture in association with membrane surfaces is probably dependent on the lipid composition and the MPER sequence range considered. The different surface hydration degrees of lipid bilayers, the hydrophobic thicknesses, and the material (elastic) properties all might restrict in an interrelated way the packing and surface exposure of different phospholipid molecules for protein binding (McIntosh & Simon, 2007; Phillips, Ursell, Wiggins, & Sens, 2009). Systematic analyses correlating changes in the bilayer material properties and immunogenicity of a given MPER peptide are to date missing. Even though, several observations indicate that individual lipids can be important to determine MPER recognition at the membrane surface. For instance, Chol, which incorporates at high concentrations into the viral envelope (ca. 45 mol%; Brugger et al., 2006), may enhance binding by 4E10. The MPER peptide preferentially inserts at the level of the acyl chain or head-group regions, with low and high Chol concentrations, respectively (Ivankin, Apellaniz, Gidalevitz, & Nieva, 2012). Thus, the increase in phospholipid cohesion induced by Chol seems to lift 4E10 epitope from the bilayer interior into the surface for better binding to the antibody.

5. IMMUNOGENICITY STUDIES OF MPER-DERIVED PEPTIDES IN LIPOSOMES

As previously discussed, lipid bilayers may provide the adequate substrate for MPER–antibody interactions, and therefore be required for recovering bNAbs combining the capacity for specific binding to MPER sequences and partitioning into membranes. Thus, in search of effective anti-MPER immunogenic responses, liposomes are not just passive adjuvant agents, but active players that shape relevant structures of MPER on their surfaces and select for distinct antibody-binding modes. Table 2 summarizes reported immunization assays based on liposomes and MPER constructs. In what follows, we will describe some attempts made in the last decade involving liposome–peptide formulations.

One of the first researchers to combine liposome and adjuvant technology with the development of MPER-targeted peptide vaccines has been C.R. Alving (Alving et al., 2012). Liposomes containing PC, Chol and PG, and GalCer or PIP, a long version of the MPER peptide including part of CHR domain and MPLA as an adjuvant, induced weak serum IgG titers in immunized mice (Karasavvas et al., 2008). Monoclonal IgM isolated from these sera revealed that the immune response targeted the immunogen but did not recognize a shorter version of the MPER including the 2F5 and 4E10/10E8 MAb epitopes (Beck, Karasavvas, Matyas, & Alving, 2008; Karasavvas et al., 2008). Interestingly, though, one of the IgMs produced did not recognize the lipids or the peptide included in the formulation when each was presented as a separate capture antigen in ELISA but, in contrast, it did recognize the peptide presented in the original immunizing liposomal formulation in the absence of PIP. The authors argued that this membrane-specific binding would arise from conformation patterns resulting from the combination of both components in the vaccine (Beck et al., 2008). In their subsequent report, the same PIP-containing MPLA-adjuvanted liposomes were doped with a shorter version of the MPER (Matyas et al., 2009). Although weak as in the previous reports, this time humoral response was focused to the MPER domain: two monoclonal IgM antibodies, WR316 and WR320, were purified, whose core epitopes overlapped that of the 4E10 or included that of the 2F5, respectively. Moreover, both were shown to bind to lipids, especially PIP and PIP2 and, importantly, the 2F5 epitope recognizing WR320 antibody weakly neutralized viral infection of PBMC. Therefore, these findings support the idea that avoidance of immunodominant sequences by designing smaller peptides representing the relevant epitopes, and combination of these peptides with membranes, might be of relevance for eliciting antibodies that target the membrane-interacting MPER by vaccination.

A report comparing the structure and immunogenicity of an MPER surrogate peptide (aa 656–683) formulated in different environments further supported a role of the membrane in shaping the conformation of peptide antigens (Serrano et al., 2014). This peptide included an aromatic-rich block allowing its insertion into the membrane interface of the liposomes (aa 678–682), a region proposed to establish weak contacts with the CDR-H3 loop of the antibody (aa 669–673) (Lorizate, Gomara, de la Torre, Andreu, & Nieva, 2006), and bear the complete 2F5 epitope (aa 656–671) (Parker et al., 2001), which folds into a continuous kinked helix in apolar environments (Serrano et al., 2014). Infrared spectroscopy confirmed the preservation of the main helical conformation in adjuvants representing licensed vaccine formulations (i.e., aluminum salt and water-in-oil emulsions) and in two different types of MPLA-adjuvanted liposomes with anionic PG or neutral PC:Chol compositions, although in the polar medium provided by alum, aggregation was more evident, as expected. The peptide was immunogenic in all four compositions (reciprocal serum endpoint titers higher than 10^3), and polyclonal 2F5 epitope-specific IgGs were recovered from the sera. This indicates, as shown also by others (Table 2; Montero et al., 2008), that properly designed and adjuvanted, MPER-derived peptides are immunogenic. However, significant neutralizing activity of the purified antibodies was only detected in PG-containing formulation. Neutralization correlated with the necessity of the unmutated CDR-H3 loop in FAb 2F5 to ensure binding to the MPER in this composition, whereas a mutant FAb 2F5 devoid of the loop still bound to the peptide in PC:Chol-containing vesicles. Therefore, antibodies raised by the former were more 2F5 alike. Moreover, molecular dynamics simulations revealed different organization of the peptides in the membranes composed of neutral or anionic compositions what might be related to the different features of the elicited antibodies. In addition, in a previous report by the same group, an alternative presentation of the MPER in the neutral Chol-containing membranes was assayed (Huarte et al., 2012). In that case, a noncovalent complex between the MPER and the FP was chosen as peptide immunogen, since previous work had suggested that amino-terminal gp41 FP residues might stabilize the 2F5 core-epitope nonhelical structure (de la Arada et al., 2009; Julien et al., 2008; Lorizate, de la Arada, et al., 2006; Lorizate, Gomara, et al., 2006). In neutral liposomes plus muramyldipeptide as an adjuvant, this complex elicited neutralizing antibodies in immunized rabbit serum (Huarte et al., 2012), which were not generated in formulations of the complex in Alum or Freund's adjuvant (Lorizate, Gomara, et al., 2006). 4E10-like responses elicited by these formulations were not reported, nor by alternative formulations assayed by other authors provided that specificity was mapped (Table 2). Interestingly, there are a few exceptions in which a weak anti-4E10 targeted response is elicited, that involve covalent conjugation of membrane anchors to the MPER (Kim et al., 2013; Watson & Szoka, 2009). The importance of the membrane anchor is further emphasized by the finding that the gp41 TM residues play a pivotal role in orienting the epitope recognized by the 4E10 antibody (Montero et al., 2012). Thus, the results of these reports indicate that both the peptide immunogen and the lipid composition, and the appropriate combination of the two need to be optimized in order to elicit a targeted and functional immune response.

Thus, another strategy adopted by the groups of Szoka and Reinherz, explores the effect of covalent attachment of lipid anchors to the MPER on the enhancement of the humoral immune response of the peptides presented in liposomes (Hanson et al., 2015; Kim et al., 2013; Venditto et al., 2013; Watson, Platt, Cao, Venditto, & Szoka, 2011; Watson & Szoka, 2009). In one report, an MPER peptide bearing the complete 2F5 epitope but devoid of the C-terminal residues (N-MPER) was conjugated at its C-terminus to a series of lipids representing some general lipid types (Watson & Szoka, 2009). As described in Table 2, except PC and palmitoyl anchors (groups 2 and 3), the other lipids attached to the peptide (groups 1 and 4–8) enhanced serum anti-MPER IgG response in immunized BALB/c mice, requiring only two immunizations to be elicited. From all the lipids assayed, conjugation to cholesterol hemisuccinate (CHEMS) was the most successful, provided MPLA was present (group 1). An unconjugated version of the peptide (group 9) in liposomes yielded almost no signal, probably due to the fact that the peptide sequence did not include the membrane-interacting aromatic-rich C-terminal region. In contrast, an unconjugated C-terminally extended version of the peptide (NC-MPER) elicited a high IgG response mapping to the 2F5 epitope (group 17). This extended peptide was also conjugated to CHEMS at its C-terminus (group 15) what improved the titer further. However, conjugation at the N-terminus or in both ends (groups 18 and 19) reduced the titer 30-fold compared to the unconjugated version. This result might be indicative of a reduced surface exposure of the N-terminal 2F5 epitope in the liposomes due to conjugation.

In addition, N-, C-, or NC-CHEMS conjugation of this extended peptide and C-CHEMS conjugation of a peptide representing the C-terminal region of the MPER (C-MPER, aa 671–683) bearing the 4E10 epitope, elicited a weak humoral response against this region. In a later contribution, noncovalent attachment of the N-MPER to liposomes was shown to improve the immunogenicity of the formulation 10-fold compared with the free peptide in liposomes (Watson et al., 2011). However, CHEMS covalent attachment still proved to be superior. Finally, in an attempt to improve the immunogenicity and to test if amino acid side chain modifications that emulate hydrophilic phospholipid head groups could contribute to the generation of 2F5 or 4E10-like neutralizing antibodies, a series of chemically modified MPER-CHEMS peptides were synthesized through derivatization of amino acid side chains with phosphate or nitrate groups (Venditto et al., 2013). These authors found that an immunogen containing both the 2F5 and 4E10 epitopes and a phosphorylated threonine, Thr-676 (Table 2, group 2) elicited higher IgG titers than its unmodified version (group 1), although unfortunately, the serum lacked detectable neutralizing activity. In contrast, modifications in other residues reduced the immunogenicity of the peptides.

Thus, residue-tuning and covalent attachment of CHEMS to peptides representing extended versions of the MPER (aa 656–683), seem to enhance the humoral response when formulated in adjuvanted liposomes (compare groups 1 vs. 17, 17 vs. 15, and 14 vs. 10, Watson & Szoka, 2009; and groups 1 vs. 2 Venditto et al., 2013; Table 2). Systematic analysis of the lipid composition in the carrier liposomes might be another interesting factor affecting the CHEMS-conjugated NC-MPER-induced humoral response and should be analyzed.

In yet another example of covalent modification of the MPER to enhance its attachment to liposomes, Reinherz and coworkers analyzed the effect of alkylation on an MPER peptide (aa 662–683) (Kim et al., 2013). Alkylated versions were shown to be more immunogenic than the free peptide in adjuvanted liposomes (Table 2, groups 1 and 2) what might be correlated to the dissociation of the unconjugated peptide from the liposomes in the lymph nodes. Although Watson and coworkers had shown that C-terminal palmitoylation of their N-MPER peptide abrogated immunogenicity (Watson & Szoka, 2009), the MPER and MPERkk-C-palm versions engineered by Kim and coworkers yielded positive results targeting the C-terminal region of the MPER (Kim et al., 2013). Subtle differences in the peptide sequences, liposome composition, and adjuvants employed might account for the conflicting results. MPERkk-C-palm was shown to be more immunogenic than MPER despite deeper immersion into the membrane acyl chain region.

Similarly, N-terminal palmitoylation yielded IgG antibodies targeting the C-terminal region of the MPER. The resulting peptide was used as a template to carry out a comprehensive characterization of the several components present in the vaccine:

- First, regarding the peptide itself, amino acid spacers of increasing length were introduced between the lipid anchor and the MPER peptide to try to switch on the immunogenicity of the N-terminal region. The length of the spacer affected 2F5, but not 4E10 binding, suggesting that the structural constraints imposed by conjugation at the MPER N-terminal region are reduced. However, immunization by this peptide in several formulations did not raise an antibody response targeting the 2F5 epitope containing N-terminal region, although mutation of the immunodominant W680 residue or substitution of the C-terminal palmitoylation by the inclusion of gp41 TM (aa 662–705) shifted the immune focus toward the N-terminal region of MPER. This indicates that provided the overall immunogenic structure and orientation of the MPER is preserved, mutation of immunodominant regions or employment of alternative membrane anchors might modulate the specificity of the immune response.
- Second, the inclusion of PEGylated phospholipid in the liposomes was shown not to affect the antigenicity or immunogenicity of the formulations (Hanson et al., 2015; Kim et al., 2013).
- Third, although immunization with liposomes carrying lipid-anchored MPER and helper peptides without additional molecular adjuvants elicited a very weak immune response, the active role of the liposomes as adjuvants was demonstrated, since formulation of MPLA and Npalm-MPER in alum or oil-based emulsions did not elicit a consistent immune response (Hanson et al., 2015). Moreover, it was found that the same outcome could be obtained by substitution of MPLA in the liposomes by TLR9 or STING agonists (Hanson et al., 2015).
- Fourth, HIV-30, a gp120-derived peptide T helper peptide proved superior than the H-2d-restricted LACK1 peptide (Hanson et al., 2015) and at the same time, the latter was more active than the universal T-cell epitope PADRE, both encapsulated in the vesicles (Kim et al., 2013).
- Fifth, a very low fluidity (DSPG:DSPC) of the liposomal phospholipids abrogated immunogenicity while medium fluidity (DMPC:DOPC:DOPG) or Chol inclusion slightly enhanced it compared to the fully fluid composition represented by DOPG:DOPC (Hanson et al., 2015; Kim et al., 2013).

– Sixth, 150 nm vesicles containing 1:500 peptide-to-lipid ratio (7–17 nm distance between the peptides) were found to be optimal (Hanson et al., 2015).

Unfortunately, serum samples or IgG purified from the sera failed to neutralize viral infection (Hanson et al., 2015; Kim et al., 2013).

Using an oligomeric gp140 protein prime-unconjugated MPER-liposome boost approach, Haynes and colleagues, also tried to improve the immunogenicity of this HIV-1 gp41 region (Table 2; Dennison et al., 2011; Verkoczy et al., 2013; Yang et al., 2013; Zhang et al., 2014). A gp140 protein prime alone, adjuvanted with Emulsigen/oCpG in guinea pigs or AS01 in rhesus macaques, only elicited a weak Trp-666 specific anti-MPER response in the former model (Table 2, group 1; Dennison et al., 2011). By comparison, MPER (aa 652–671) boost alone, including TLR4, 7/8, and 9 agonists in the liposomal carrier, elicited a higher humoral response specific for the 2F5 core-epitope ^{664}DK^{665} dipeptide in one out of three guinea pigs (group 2) and specific for Asp-664 in rhesus macaques (group 7). Combination of the two was shown to further enhance the titer and focus the specificity (groups 3 and 4). Moreover, inclusion of the C-terminal residues of the MPER and two rounds of prime-boosting focused the humoral response to the 2F5 core-epitope residues 664DKW in one out of three guinea pigs and in three out of four rhesus macaques (groups 5 and 8). In addition, the elicited antibodies recognized a gp41 fusion-intermediate representing protein and were shown to bind to MPER in a conformation similar to that bound by the 2F5 MAb. These results supported the validity of the prime-boost strategy and the chosen immunogens and implied that several boostings are critical for the expansion and maturation of B-cell specificity. However, once again, serum-neutralizing activity against HIV-1 pseudoviruses was not detected, a fact that the authors related to tolerance mechanisms preventing the development of the autoreactive anti-MPER antibodies or to a possible lack of a 2F5-like CDR-H3 loop in the elicited antibodies, essential for neutralization (Dennison et al., 2011).

Indeed, the same authors identified human kynureninase (KYNU) (ELDKWA) and splicing factor 3b subunit 3 (SF3B3) as the self-antigens recognized by the 2F5 and 4E10 antibodies, respectively (Yang et al., 2013). Opossums carry a D/E replacement in the KYNU sequence that destroys the 2F5 nominal epitope and abolishes 2F5 binding, but they retain the SF3B3 4E10 epitope. Thus, it was hypothesized that, if autorreactivity were the cause of the poor immunogenicity of the MPER 2F5 epitope, an immunogen that properly represented this epitope would yield a high humoral response toward this region in this animal model. And in fact, immunization of opossums with the same immunogens used to immunize rhesus macaques in the previous report (Dennison et al., 2011) elicited extraordinary titers of serum antibodies specific for the 2F5 core epitope amino acid Asp-664, but not to the 4E10 epitope. The absence of response toward the 4E10 epitope was also ascribed to autoreactivity; however, suboptimal representation of the 4E10 epitope by the chosen immunogen cannot be ruled out. Despite the high titers obtained, neutralization activity was again absent (Yang et al., 2013).

Based on these findings, a similar immunogen (membrane-anchored MPER) and a similar immunizing strategy was selected to demonstrate that robust serum-neutralizing IgG responses might be induced if, after tolerance, residual 2F5-expressing B-cells still present in 2F5 knockin mice are activated by a structurally compatible immunogen (Verkoczy et al., 2013). Furthermore, this study demonstrates that all three components in the immunization regimen (i.e., TLR agonists, MPER peptides, and liposomes) are required for triggering and expanding preexisting, anergic 2F5-expressing B-cells (Table 2).

Finally, characterization the effect of host genetics on the humoral response to the immunogens revealed that the induction of rapid, sustained, and elevated anti-MPER antibody titers largely depends on the MHC-II haplotype-restricted CD4 T_H cell activation in mice (Zhang et al., 2014). Prime-boost vaccination with membrane-anchored MPER elicited outstanding, ^{665}KW^{666} specific, nonneutralizing IgG titers in H-2^d restricted BALB/c mice. In contrast, H-2^b- or H-2^k-restricted mice elicited 4–5 log lower titers regardless of the adjuvants and HIV-1 immunogens used. Moreover, a humoral response against the immunogen could not be detected in H-2^s-restricted mice (Table 2). The core TH epitope for H-2^d restriction was found to be KWASLWNWF partially overlapping the epitopes of the known anti-MPER antibodies. This might be disadvantageous for B-cell immunogenicity since competition between BCR and class II molecules for antigen binding might occur (Zhang et al., 2014).

All in all, the immunization studies analyzed in detail in this section and other reports summarized in Table 2 indicate that all the components in the vaccine and genetic factors have to be systematically characterized and rationally combined to elicit the longed-for neutralizing antibody response.

REFERENCES

Alving, C. R. (2008). 4E10 and 2F5 monoclonal antibodies: Binding specificities to phospholipids, tolerance, and clinical safety issues. *AIDS, 22*(5), 649–651.

Alving, C. R., & Rao, M. (2008). Lipid A and liposomes containing lipid A as antigens and adjuvants. *Vaccine, 26*(24), 3036–3045.

Alving, C. R., Rao, M., Steers, N. J., Matyas, G. R., & Mayorov, A. V. (2012). Liposomes containing lipid A: An effective, safe, generic adjuvant system for synthetic vaccines. *Expert Review of Vaccines, 11*(6), 733–744.

Barbato, G., Bianchi, E., Ingallinella, P., Hurni, W. H., Miller, M. D., Ciliberto, G., et al. (2003). Structural analysis of the epitope of the anti-HIV antibody 2F5 sheds light into its mechanism of neutralization and HIV fusion. *Journal of Molecular Biology, 330*(5), 1101–1115.

Barouch, D. H. (2007). HIV-1 vaccine candidate ineffective. *AIDS Clinical Care, 19*(11), 93.

Beck, Z., Karasavvas, N., Matyas, G. R., & Alving, C. R. (2008). Membrane-specific antibodies induced by liposomes can simultaneously bind to HIV-1 protein, peptide, and membrane lipid epitopes. *Journal of Drug Targeting, 16*(7), 535–542.

Benen, T. D., Tonks, P., Kliche, A., Kapzan, R., Heeney, J. L., & Wagner, R. (2014). Development and immunological assessment of VLP-based immunogens exposing the membrane-proximal region of the HIV-1 gp41 protein. *Journal of Biomedical Science, 21*, 79.

Binley, J. M., Wrin, T., Korber, B., Zwick, M. B., Wang, M., Chappey, C., et al. (2004). Comprehensive cross-clade neutralization analysis of a panel of anti-human immunodeficiency virus type 1 monoclonal antibodies. *Journal of Virology, 78*(23), 13232–13252.

Bird, G. H., Irimia, A., Ofek, G., Kwong, P. D., Wilson, I. A., & Walensky, L. D. (2014). Stapled HIV-1 peptides recapitulate antigenic structures and engage broadly neutralizing antibodies. *Nature Structural & Molecular Biology, 21*(12), 1058–1067.

Biron, Z., Khare, S., Samson, A. O., Hayek, Y., Naider, F., & Anglister, J. (2002). A monomeric 3(10)-helix is formed in water by a 13-residue peptide representing the neutralizing determinant of HIV-1 on gp41. *Biochemistry, 41*(42), 12687–12696.

Bittle, J. L., Houghten, R. A., Alexander, H., Shinnick, T. M., Sutcliffe, J. G., Lerner, R. A., et al. (1982). Protection against foot-and-mouth disease by immunization with a chemically synthesized peptide predicted from the viral nucleotide sequence. *Nature, 298*(5869), 30–33.

Black, M., Trent, A., Tirrell, M., & Olive, C. (2010). Advances in the design and delivery of peptide subunit vaccines with a focus on toll-like receptor agonists. *Expert Review of Vaccines, 9*(2), 157–173.

Blattner, C., Lee, J. H., Sliepen, K., Derking, R., Falkowska, E., de la Pena, A. T., et al. (2014). Structural delineation of a quaternary, cleavage-dependent epitope at the gp41-gp120 interface on intact HIV-1 Env trimers. *Immunity, 40*(5), 669–680.

Brugger, B., Glass, B., Haberkant, P., Leibrecht, I., Wieland, F. T., & Krausslich, H. G. (2006). The HIV lipidome: A raft with an unusual composition. *Proceedings of the National Academy of Sciences of the United States of America, 103*(8), 2641–2646.

Brunel, F. M., Zwick, M. B., Cardoso, R. M., Nelson, J. D., Wilson, I. A., Burton, D. R., et al. (2006). Structure-function analysis of the epitope for 4E10, a broadly neutralizing human immunodeficiency virus type 1 antibody. *Journal of Virology, 80*(4), 1680–1687.

Brunner, R., Jensen-Jarolim, E., & Pali-Schöll, I. (2010). The ABC of clinical and experimental adjuvants—A brief overview. *Immunology Letters, 128*(1), 29–35.

Burton, D. R. (2010). Scaffolding to build a rational vaccine design strategy. *Proceedings of the National Academy of Sciences of the United States of America, 107*(42), 17859–17860.

Burton, D. R., Ahmed, R., Barouch, D. H., Butera, S. T., Crotty, S., Godzik, A., et al. (2012). A blueprint for HIV vaccine discovery. *Cell Host & Microbe, 12*(4), 396–407.

Burton, D. R., Stanfield, R. L., & Wilson, I. A. (2005). Antibody vs. HIV in a clash of evolutionary titans. *Proceedings of the National Academy of Sciences of the United States of America, 102*(42), 14943–14948.

Cantor, R. S. (1999). Lipid composition and the lateral pressure profile in bilayers. *Biophysical Journal, 76*(5), 2625–2639.

Caoili, S. E. (2013). Antidotes, antibody-mediated immunity and the future of pharmaceutical product development. *Human Vaccines Immunotherapy, 9*(2), 294–299.

Cardoso, R. M., Brunel, F. M., Ferguson, S., Zwick, M., Burton, D. R., Dawson, P. E., et al. (2007). Structural basis of enhanced binding of extended and helically constrained peptide epitopes of the broadly neutralizing HIV-1 antibody 4E10. *Journal of Molecular Biology, 365*(5), 1533–1544.

Cardoso, R. M., Zwick, M. B., Stanfield, R. L., Kunert, R., Binley, J. M., Katinger, H., et al. (2005). Broadly neutralizing anti-HIV antibody 4E10 recognizes a helical conformation of a highly conserved fusion-associated motif in gp41. *Immunity, 22*(2), 163–173.

Coeffier, E., Clement, J. M., Cussac, V., Khodaei-Boorane, N., Jehanno, M., Rojas, M., et al. (2000). Antigenicity and immunogenicity of the HIV-1 gp41 epitope ELDKWA inserted into permissive sites of the MalE protein. *Vaccine, 19*(7–8), 684–693.

Coffman, R. L., Sher, A., & Seder, R. A. (2010). Vaccine adjuvants: Putting innate immunity to work. *Immunity, 33*(4), 492–503.

Correia, B. E., Ban, Y. E., Holmes, M. A., Xu, H., Ellingson, K., Kraft, Z., et al. (2010). Computational design of epitope-scaffolds allows induction of antibodies specific for a poorly immunogenic HIV vaccine epitope. *Structure, 18*(9), 1116–1126.

Correia, B. E., Bates, J. T., Loomis, R. J., Baneyx, G., Carrico, C., Jardine, J. G., et al. (2014). Proof of principle for epitope-focused vaccine design. *Nature, 507*(7491), 201–206.

Crooks, G. E., Hon, G., Chandonia, J. M., & Brenner, S. E. (2004). WebLogo: A sequence logo generator. *Genome Research, 14*(6), 1188–1190.

de la Arada, I., Julien, J. P., de la Torre, B. G., Huarte, N., Andreu, D., Pai, E. F., et al. (2009). Structural constraints imposed by the conserved fusion peptide on the HIV-1 gp41 epitope recognized by the broadly neutralizing antibody 2F5. *The Journal of Physical Chemistry B, 113*(41), 13626–13637.

Dennison, S. M., Stewart, S. M., Stempel, K. C., Liao, H. X., Haynes, B. F., & Alam, S. M. (2009). Stable docking of neutralizing human immunodeficiency virus type 1 gp41 membrane-proximal external region monoclonal antibodies 2F5 and 4E10 is dependent on the membrane immersion depth of their epitope regions. *Journal of Virology, 83*(19), 10211–10223.

Dennison, S. M., Sutherland, L. L., Jaeger, F. H., Anasti, K. M., Parks, R., Stewart, S., et al. (2011). Induction of antibodies in rhesus macaques that recognize a fusion-intermediate conformation of HIV-1 gp41. *PLoS One, 6*(11), e27824.

Devito, C., Zuber, B., Schroder, U., Benthin, R., Okuda, K., Broliden, K., et al. (2004). Intranasal HIV-1-gp160-DNA/gp41 peptide prime-boost immunization regimen in mice results in long-term HIV-1 neutralizing humoral mucosal and systemic immunity. *Journal of Immunology, 173*(11), 7078–7089.

Dyson, H. J., Lerner, R. A., & Wright, P. E. (1988). The physical basis for induction of protein-reactive antipeptide antibodies. *Annual Review of Biophysics and Biophysical Chemistry, 17*, 305–324.

Esparza, J. (2013). A brief history of the global effort to develop a preventive HIV vaccine. *Vaccine, 31*(35), 3502–3518.

Falkowska, E., Le, K. M., Ramos, A., Doores, K. J., Lee, J. H., Blattner, C., et al. (2014). Broadly neutralizing HIV antibodies define a glycan-dependent epitope on the prefusion conformation of gp41 on cleaved envelope trimers. *Immunity, 40*(5), 657–668.

Ferrantelli, F., Hofmann-Lehmann, R., Rasmussen, R. A., Wang, T., Xu, W., Li, P. L., et al. (2003). Post-exposure prophylaxis with human monoclonal antibodies prevented SHIV89.6P infection or disease in neonatal macaques. *AIDS, 17*(3), 301–309.

Francis, M. J. (2003). Synthetic peptides. *Methods in Molecular Medicine*, *87*, 115–132.
Goni, F., Mathiason, C. K., Yim, L., Wong, K., Hayes-Klug, J., Nalls, A., et al. (2015). Mucosal immunization with an attenuated Salmonella vaccine partially protects white-tailed deer from chronic wasting disease. *Vaccine*, *33*, 726–733.
Gray, E. S., Madiga, M. C., Moore, P. L., Mlisana, K., Abdool Karim, S. S., Binley, J. M., et al. (2009). Broad neutralization of human immunodeficiency virus type 1 mediated by plasma antibodies against the gp41 membrane proximal external region. *Journal of Virology*, *83*(21), 11265–11274.
Guenaga, J., Dosenovic, P., Ofek, G., Baker, D., Schief, W. R., Kwong, P. D., et al. (2011). Heterologous epitope-scaffold prime:boosting immuno-focuses B cell responses to the HIV-1 gp41 2F5 neutralization determinant. *PLoS One*, *6*(1), e16074.
Hafez, I. M., & Cullis, P. R. (2000). Cholesteryl hemisuccinate exhibits pH sensitive polymorphic phase behavior. *Biochimica et Biophysica Acta (BBA) - Biomembranes*, *1463*(1), 107–114.
Hafner, A. M., Corthesy, B., & Merkle, H. P. (2013). Particulate formulations for the delivery of poly(I:C) as vaccine adjuvant. *Advanced Drug Delivery Reviews*, *65*(10), 1386–1399.
Hanson, M. C., Abraham, W., Crespo, M. P., Chen, S. H., Liu, H., Szeto, G. L., et al. (2015). Liposomal vaccines incorporating molecular adjuvants and intrastructural T-cell help promote the immunogenicity of HIV membrane-proximal external region peptides. *Vaccine*, *33*(7), 861–868.
Haynes, B. F., Fleming, J., St Clair, E. W., Katinger, H., Stiegler, G., Kunert, R., et al. (2005). Cardiolipin polyspecific autoreactivity in two broadly neutralizing HIV-1 antibodies. *Science*, *308*(5730), 1906–1908.
Ho, J., Uger, R. A., Zwick, M. B., Luscher, M. A., Barber, B. H., & MacDonald, K. S. (2005). Conformational constraints imposed on a pan-neutralizing HIV-1 antibody epitope result in increased antigenicity but not neutralizing response. *Vaccine*, *23*(13), 1559–1573.
Hope, M. J., Bally, M. B., Webb, G., & Cullis, P. R. (1985). Production of large unilamellar vesicles by a rapid extrusion procedure. Characterization of size distribution, trapped volume and ability to maintain a membrane potential. *Biochimica et Biophysica Acta*, *812*, 55–65.
Huang, J., Kang, B. H., Pancera, M., Lee, J. H., Tong, T., Feng, Y., et al. (2014). Broad and potent HIV-1 neutralization by a human antibody that binds the gp41-gp120 interface. *Nature*.
Huang, J., Ofek, G., Laub, L., Louder, M. K., Doria-Rose, N. A., Longo, N. S., et al. (2012). Broad and potent neutralization of HIV-1 by a gp41-specific human antibody. *Nature*, *491*(7424), 406–412.
Huarte, N., Araujo, A., Arranz, R., Lorizate, M., Quendler, H., Kunert, R., et al. (2012). Recognition of membrane-bound fusion-peptide/MPER complexes by the HIV-1 neutralizing 2F5 antibody: Implications for anti-2F5 immunogenicity. *PLoS One*, *7*(12), e52740.
Huarte, N., Lorizate, M., Maeso, R., Kunert, R., Arranz, R., Valpuesta, J. M., et al. (2008). The broadly neutralizing anti-human immunodeficiency virus type 1 4E10 monoclonal antibody is better adapted to membrane-bound epitope recognition and blocking than 2F5. *Journal of Virology*, *82*(18), 8986–8996.
Ingale, S., Gach, J. S., Zwick, M. B., & Dawson, P. E. (2010). Synthesis and analysis of the membrane proximal external region epitopes of HIV-1. *Journal of Peptide Science*, *16*(12), 716–722.
Ivankin, A., Apellaniz, B., Gidalevitz, D., & Nieva, J. L. (2012). Mechanism of membrane perturbation by the HIV-1 gp41 membrane-proximal external region and its modulation by cholesterol. *Biochimica et Biophysica Acta*, *1818*(11), 2521–2528.
Julien, J. P., Bryson, S., Nieva, J. L., & Pai, E. F. (2008). Structural details of HIV-1 recognition by the broadly neutralizing monoclonal antibody 2F5: Epitope conformation,

antigen-recognition loop mobility, and anion-binding site. *Journal of Molecular Biology, 384*(2), 377–392.

Julien, J. P., Cupo, A., Sok, D., Stanfield, R. L., Lyumkis, D., Deller, M. C., et al. (2013). Crystal structure of a soluble cleaved HIV-1 envelope trimer. *Science, 342*(6165), 1477–1483.

Julien, J. P., Huarte, N., Maeso, R., Taneva, S. G., Cunningham, A., Nieva, J. L., et al. (2010). Ablation of the complementarity-determining region H3 apex of the anti-HIV-1 broadly neutralizing antibody 2F5 abrogates neutralizing capacity without affecting core epitope binding. *Journal of Virology, 84*(9), 4136–4147.

Kamdem Toukam, D., Tenbusch, M., Stang, A., Temchura, V., Storcksdieck Genannt Bonsmann, M., Grewe, B., et al. (2012). Targeting antibody responses to the membrane proximal external region of the envelope glycoprotein of human immunodeficiency virus. *PLoS One*, 7(5), e38068.

Karasavvas, N., Beck, Z., Tong, J., Matyas, G. R., Rao, M., McCutchan, F. E., et al. (2008). Antibodies induced by liposomal protein exhibit dual binding to protein and lipid epitopes. *Biochemical and Biophysical Research Communications, 366*(4), 982–987.

Kersten, G. F., & Crommelin, D. J. (2003). Liposomes and ISCOMs. *Vaccine, 21*(9–10), 915–920.

Khayat, R., Lee, J. H., Julien, J. P., Cupo, A., Klasse, P. J., Sanders, R. W., et al. (2013). Structural characterization of cleaved, soluble HIV-1 envelope glycoprotein trimers. *Journal of Virology, 87*(17), 9865–9872.

Kim, A. S., Leaman, D. P., & Zwick, M. B. (2014). Antibody to gp41 MPER alters functional properties of HIV-1 Env without complete neutralization. *PLoS Pathogens, 10*(7), e1004271.

Kim, M., Song, L., Moon, J., Sun, Z. Y., Bershteyn, A., Hanson, M., et al. (2013). Immunogenicity of membrane-bound HIV-1 gp41 membrane-proximal external region (MPER) segments is dominated by residue accessibility and modulated by stereochemistry. *The Journal of Biological Chemistry, 288*(44), 31888–31901.

Klasse, P. J., Depetris, R. S., Pejchal, R., Julien, J. P., Khayat, R., Lee, J. H., et al. (2013). Influences on trimerization and aggregation of soluble, cleaved HIV-1 SOSIP envelope glycoprotein. *Journal of Virology, 87*(17), 9873–9885.

Klein, F., Diskin, R., Scheid, J. F., Gaebler, C., Mouquet, H., Georgiev, I. S., et al. (2013). Somatic mutations of the immunoglobulin framework are generally required for broad and potent HIV-1 neutralization. *Cell, 153*(1), 126–138.

Klein, F., Mouquet, H., Dosenovic, P., Scheid, J. F., Scharf, L., & Nussenzweig, M. C. (2013). Antibodies in HIV-1 vaccine development and therapy. *Science, 341*(6151), 1199–1204.

Kulp, D. W., & Schief, W. R. (2013). Advances in structure-based vaccine design. *Current Opinion in Virology, 3*(3), 322–331.

Kunert, R., Ruker, F., & Katinger, H. (1998). Molecular characterization of five neutralizing anti-HIV type 1 antibodies: Identification of nonconventional D segments in the human monoclonal antibodies 2G12 and 2F5. *AIDS Research and Human Retroviruses, 14*(13), 1115–1128.

Kwong, P. D., Doyle, M. L., Casper, D. J., Cicala, C., Leavitt, S. A., Majeed, S., et al. (2002). HIV-1 evades antibody-mediated neutralization through conformational masking of receptor-binding sites. *Nature, 420*(6916), 678–682.

Kwong, P. D., & Mascola, J. R. (2012). Human antibodies that neutralize HIV-1: Identification, structures, and B cell ontogenies. *Immunity, 37*(3), 412–425.

Lai, R. P., Hock, M., Radzimanowski, J., Tonks, P., Hulsik, D. L., Effantin, G., et al. (2014). A fusion intermediate gp41 immunogen elicits neutralizing antibodies to HIV-1. *The Journal of Biological Chemistry, 289*(43), 29912–29926.

Langeveld, J. P., Casal, J. I., Cortes, E., van de Wetering, G., Boshuizen, R. S., Schaaper, W. M., et al. (1994). Effective induction of neutralizing antibodies with the amino terminus of VP2 of canine parvovirus as a synthetic peptide. *Vaccine*, *12*(15), 1473–1480.

Lee, B. S., Huang, J. S., Jayathilaka, G. D., Lateef, S. S., & Gupta, S. (2010). Production of antipeptide antibodies. *Methods in Molecular Biology*, *657*, 93–108.

Lenz, O., Dittmar, M. T., Wagner, A., Ferko, B., Vorauer-Uhl, K., Stiegler, G., et al. (2005). Trimeric membrane-anchored gp41 inhibits HIV membrane fusion. *The Journal of Biological Chemistry*, *280*(6), 4095–4101.

Li, W., Joshi, M., Singhania, S., Ramsey, K., & Murthy, A. (2014). Peptide vaccine: Progress and challenges. *Vaccines*, *2*(3), 515–536.

Li, Y., Svehla, K., Louder, M. K., Wycuff, D., Phogat, S., Tang, M., et al. (2009). Analysis of neutralization specificities in polyclonal sera derived from human immunodeficiency virus type 1-infected individuals. *Journal of Virology*, *83*(2), 1045–1059.

Lorizate, M., Cruz, A., Huarte, N., Kunert, R., Perez-Gil, J., & Nieva, J. L. (2006). Recognition and blocking of HIV-1 gp41 pre-transmembrane sequence by monoclonal 4E10 antibody in a Raft-like membrane environment. *The Journal of Biological Chemistry*, *281*(51), 39598–39606.

Lorizate, M., de la Arada, I., Huarte, N., Sanchez-Martinez, S., de la Torre, B. G., Andreu, D., et al. (2006). Structural analysis and assembly of the HIV-1 Gp41 amino-terminal fusion peptide and the pretransmembrane amphipathic-at-interface sequence. *Biochemistry*, *45*(48), 14337–14346.

Lorizate, M., Gomara, M. J., de la Torre, B. G., Andreu, D., & Nieva, J. L. (2006). Membrane-transferring sequences of the HIV-1 Gp41 ectodomain assemble into an immunogenic complex. *Journal of Molecular Biology*, *360*(1), 45–55.

Lorizate, M., Huarte, N., Saez-Cirion, A., & Nieva, J. L. (2008). Interfacial pre-transmembrane domains in viral proteins promoting membrane fusion and fission. *Biochimica et Biophysica Acta*, *1778*, 1624–1639.

Lutje Hulsik, D., Liu, Y. Y., Strokappe, N. M., Battella, S., El Khattabi, M., McCoy, L. E., et al. (2013). A gp41 MPER-specific llama VHH requires a hydrophobic CDR3 for neutralization but not for antigen recognition. *PLoS Pathogens*, *9*(3), e1003202.

Lyumkis, D., Julien, J. P., de Val, N., Cupo, A., Potter, C. S., Klasse, P. J., et al. (2013). Cryo-EM structure of a fully glycosylated soluble cleaved HIV-1 envelope trimer. *Science*, *342*(6165), 1484–1490.

Maeso, R., Huarte, N., Julien, J. P., Kunert, R., Pai, E. F., & Nieva, J. L. (2011). Interaction of anti-HIV type 1 antibody 2F5 with phospholipid bilayers and its relevance for the mechanism of virus neutralization. *AIDS Research and Human Retroviruses*, *27*(8), 863–876.

Matyas, G. R., Wieczorek, L., Beck, Z., Ochsenbauer-Jambor, C., Kappes, J. C., Michael, N. L., et al. (2009). Neutralizing antibodies induced by liposomal HIV-1 glycoprotein 41 peptide simultaneously bind to both the 2F5 or 4E10 epitope and lipid epitopes. *AIDS*, *23*(16), 2069–2077.

McGaughey, G. B., Citron, M., Danzeisen, R. C., Freidinger, R. M., Garsky, V. M., Hurni, W. M., et al. (2003). HIV-1 vaccine development: Constrained peptide immunogens show improved binding to the anti-HIV-1 gp41 MAb. *Biochemistry*, *42*(11), 3214–3223.

McIntosh, T. J., & Simon, S. A. (2007). Bilayers as protein solvents: Role of bilayer structure and elastic properties. *The Journal of General Physiology*, *130*(2), 225–227.

McLellan, J. S., Chen, M., Joyce, M. G., Sastry, M., Stewart-Jones, G. B., Yang, Y., et al. (2013). Structure-based design of a fusion glycoprotein vaccine for respiratory syncytial virus. *Science*, *342*(6158), 592–598.

Mohan, T., Verma, P., & Rao, D. N. (2014). Comparative mucosal immunogenicity of HIV gp41 membrane-proximal external region (MPER) containing single and multiple repeats of ELDKWA sequence with defensin peptides. *Immunobiology*, *219*(4), 292–301.

Moisa, A. A., & Kolesanova, E. F. (2011). Synthetic peptide vaccines. *Biomeditsinskaya Khimiya*, *57*(1), 14–30.
Montero, M., Gulzar, N., Klaric, K. A., Donald, J. E., Lepik, C., Wu, S., et al. (2012). Neutralizing epitopes in the membrane-proximal external region of HIV-1 gp41 are influenced by the transmembrane domain and the plasma membrane. *Journal of Virology*, *86*(6), 2930–2941.
Montero, M., van Houten, N. E., Wang, X., & Scott, J. K. (2008). The membrane-proximal external region of the human immunodeficiency virus type 1 envelope: Dominant site of antibody neutralization and target for vaccine design. *Microbiology and Molecular Biology Reviews*, *72*(1), 54–84, table of contents.
Morris, L., Chen, X., Alam, M., Tomaras, G., Zhang, R., Marshall, D. J., et al. (2011). Isolation of a human anti-HIV gp41 membrane proximal region neutralizing antibody by antigen-specific single B cell sorting. *PLoS One*, *6*(9), e23532.
Muster, T., Guinea, R., Trkola, A., Purtscher, M., Klima, A., Steindl, F., et al. (1994). Cross-neutralizing activity against divergent human immunodeficiency virus type 1 isolates induced by the gp41 sequence ELDKWAS. *Journal of Virology*, *68*(6), 4031–4034.
Muster, T., Steindl, F., Purtscher, M., Trkola, A., Klima, A., Himmler, G., et al. (1993). A conserved neutralizing epitope on gp41 of human immunodeficiency virus type 1. *Journal of Virology*, *67*(11), 6642–6647.
Ofek, G., Guenaga, F. J., Schief, W. R., Skinner, J., Baker, D., Wyatt, R., et al. (2010). Elicitation of structure-specific antibodies by epitope scaffolds. *Proceedings of the National Academy of Sciences of the United States of America*, *107*(42), 17880–17887.
Ofek, G., McKee, K., Yang, Y., Yang, Z. Y., Skinner, J., Guenaga, F. J., et al. (2010). Relationship between antibody 2F5 neutralization of HIV-1 and hydrophobicity of its heavy chain third complementarity-determining region. *Journal of Virology*, *84*(6), 2955–2962.
Ofek, G., Tang, M., Sambor, A., Katinger, H., Mascola, J. R., Wyatt, R., et al. (2004). Structure and mechanistic analysis of the anti-human immunodeficiency virus type 1 antibody 2F5 in complex with its gp41 epitope. *Journal of Virology*, *78*(19), 10724–10737.
Ofek, G., Zirkle, B., Yang, Y., Zhu, Z., McKee, K., Zhang, B., et al. (2014). Structural basis for HIV-1 neutralization by 2F5-like antibodies m66 and m66.6. *Journal of Virology*, *88*(5), 2426–2441.
Pancera, M., Zhou, T., Druz, A., Georgiev, I. S., Soto, C., Gorman, J., et al. (2014). Structure and immune recognition of trimeric pre-fusion HIV-1 Env. *Nature*, *514*(7523), 455–461.
Parker, C. E., Deterding, L. J., Hager-Braun, C., Binley, J. M., Schulke, N., Katinger, H., et al. (2001). Fine definition of the epitope on the gp41 glycoprotein of human immunodeficiency virus type 1 for the neutralizing monoclonal antibody 2 F5. *Journal of Virology*, *75*(22), 10906–10911.
Pejchal, R., Gach, J. S., Brunel, F. M., Cardoso, R. M., Stanfield, R. L., Dawson, P. E., et al. (2009). A conformational switch in human immunodeficiency virus gp41 revealed by the structures of overlapping epitopes recognized by neutralizing antibodies. *Journal of Virology*, *83*(17), 8451–8462.
Phillips, R., Ursell, T., Wiggins, P., & Sens, P. (2009). Emerging roles for lipids in shaping membrane-protein function. *Nature*, *459*(7245), 379–385.
Plotkin, S. (2014). History of vaccination. *Proceedings of the National Academy of Sciences of the United States of America*, *111*(34), 12283–12287.
Purcell, A. W., McCluskey, J., & Rossjohn, J. (2007). More than one reason to rethink the use of peptides in vaccine design. *Nature Reviews. Drug Discovery*, *6*(5), 404–414.
Purtscher, M., Trkola, A., Gruber, G., Buchacher, A., Predl, R., Steindl, F., et al. (1994). A broadly neutralizing human monoclonal antibody against gp41 of human immunodeficiency virus type 1. *AIDS Research and Human Retroviruses*, *10*(12), 1651–1658.

Reardon, P. N., Sage, H., Dennison, S. M., Martin, J. W., Donald, B. R., Alam, S. M., et al. (2014). Structure of an HIV-1-neutralizing antibody target, the lipid-bound gp41 envelope membrane proximal region trimer. *Proceedings of the National Academy of Sciences of the United States of America, 111*(4), 1391–1396.

Rosendahl Huber, S., van Beek, J., de Jonge, J., Luytjes, W., & van Baarle, D. (2014). T cell responses to viral infections—Opportunities for peptide vaccination. *Frontiers in Immunology, 5.*

Saez-Cirion, A., Arrondo, J. L., Gomara, M. J., Lorizate, M., Iloro, I., Melikyan, G., et al. (2003). Structural and functional roles of HIV-1 gp41 pretransmembrane sequence segmentation. *Biophysical Journal, 85*(6), 3769–3780.

Santiveri, C. M., Rico, M., & Jimenez, M. A. (2000). Position effect of cross-strand side-chain interactions on beta-hairpin formation. *Protein Science, 9*(11), 2151–2160.

Santiveri, C. M., Rico, M., Jimenez, M. A., Pastor, M. T., & Perez-Paya, E. (2003). Insights into the determinants of beta-sheet stability: 1H and 13C NMR conformational investigation of three-stranded antiparallel beta-sheet-forming peptides. *The Journal of Peptide Research, 61*(4), 177–188.

Santiveri, C. M., Santoro, J., Rico, M., & Jimenez, M. A. (2004). Factors involved in the stability of isolated beta-sheets: Turn sequence, beta-sheet twisting, and hydrophobic surface burial. *Protein Science, 13*(4), 1134–1147.

Sather, D. N., & Stamatatos, L. (2010). Epitope specificities of broadly neutralizing plasmas from HIV-1 infected subjects. *Vaccine, 28*(Suppl. 2), B8–B12.

Scharf, L., Scheid, J. F., Lee, J. H., West, A. P., Jr., Chen, C., Gao, H., et al. (2014). Antibody 8ANC195 reveals a site of broad vulnerability on the HIV-1 envelope spike. *Cell Reports,* 7(3), 785–795.

Scherer, E. M., Leaman, D. P., Zwick, M. B., McMichael, A. J., & Burton, D. R. (2010). Aromatic residues at the edge of the antibody combining site facilitate viral glycoprotein recognition through membrane interactions. *Proceedings of the National Academy of Sciences of the United States of America, 107*(4), 1529–1534.

Scherer, E. M., Zwick, M. B., Teyton, L., & Burton, D. R. (2007). Difficulties in eliciting broadly neutralizing anti-HIV antibodies are not explained by cardiolipin autoreactivity. *AIDS, 21*(16), 2131–2139.

Schibli, D. J., Montelaro, R. C., & Vogel, H. J. (2001). The membrane-proximal tryptophan-rich region of the HIV glycoprotein, gp41, forms a well-defined helix in dodecylphosphocholine micelles. *Biochemistry, 40*(32), 9570–9578.

Serrano, S., Araujo, A., Apellaniz, B., Bryson, S., Carravilla, P., de la Arada, I., et al. (2014). Structure and immunogenicity of a peptide vaccine, including the complete HIV-1 gp41 2 F5 epitope: Implications for antibody recognition mechanism and immunogen design. *The Journal of Biological Chemistry, 289*(10), 6565–6580.

Sesardic, D. (1993). Synthetic peptide vaccines. *Journal of Medical Microbiology, 39*(4), 241–242.

Shen, X., Parks, R. J., Montefiori, D. C., Kirchherr, J. L., Keele, B. F., Decker, J. M., et al. (2009). In vivo gp41 antibodies targeting the 2F5 monoclonal antibody epitope mediate human immunodeficiency virus type 1 neutralization breadth. *Journal of Virology, 83*(8), 3617–3625.

Slifka, M. K., & Amanna, I. (2014). How advances in immunology provide insight into improving vaccine efficacy. *Vaccine, 32*(25), 2948–2957.

Song, L., Sun, Z. Y., Coleman, K. E., Zwick, M. B., Gach, J. S., Wang, J. H., et al. (2009). Broadly neutralizing anti-HIV-1 antibodies disrupt a hinge-related function of gp41 at the membrane interface. *Proceedings of the National Academy of Sciences of the United States of America, 106*(22), 9057–9062.

Stamatatos, L., Morris, L., Burton, D. R., & Mascola, J. R. (2009). Neutralizing antibodies generated during natural HIV-1 infection: Good news for an HIV-1 vaccine? *Nature Medicine, 15*(8), 866–870.

Stanfield, R. L., Julien, J.-P., Pejchal, R., Gach, J. S., Zwick, M. B., & Wilson, I. A. (2011). Structure-based design of a protein immunogen that displays an HIV-1 gp41 neutralizing epitope. *Journal of Molecular Biology*, *414*(3), 460–476.

Stiegler, G., Armbruster, C., Vcelar, B., Stoiber, H., Kunert, R., Michael, N. L., et al. (2002). Antiviral activity of the neutralizing antibodies 2F5 and 2G12 in asymptomatic HIV-1-infected humans: A phase I evaluation. *AIDS*, *16*(15), 2019–2025.

Stiegler, G., Kunert, R., Purtscher, M., Wolbank, S., Voglauer, R., Steindl, F., et al. (2001). A potent cross-clade neutralizing human monoclonal antibody against a novel epitope on gp41 of human immunodeficiency virus type 1. *AIDS Research and Human Retroviruses*, *17*(18), 1757–1765.

Sun, Z. Y., Cheng, Y., Kim, M., Song, L., Choi, J., Kudahl, U. J., et al. (2014). Disruption of helix-capping residues 671 and 674 reveals a role in HIV-1 entry for a specialized hinge segment of the membrane proximal external region of gp41. *Journal of Molecular Biology*, *426*(5), 1095–1108.

Sun, Z. Y., Oh, K. J., Kim, M., Yu, J., Brusic, V., Song, L., et al. (2008). HIV-1 broadly neutralizing antibody extracts its epitope from a kinked gp41 ectodomain region on the viral membrane. *Immunity*, *28*(1), 52–63.

Testa, J. S., & Philip, R. (2012). Role of T-cell epitope-based vaccine in prophylactic and therapeutic applications. *Future Virology*, 7(11), 1077–1088.

Tian, Y., Ramesh, C. V., Ma, X., Naqvi, S., Patel, T., Cenizal, T., et al. (2002). Structure-affinity relationships in the gp41 ELDKWA epitope for the HIV-1 neutralizing monoclonal antibody 2F5: Effects of side-chain and backbone modifications and conformational constraints. *The Journal of Peptide Research*, *59*(6), 264–276.

Tobin, G. J., Trujillo, J. D., Bushnell, R. V., Lin, G., Chaudhuri, A. R., Long, J., et al. (2008). Deceptive imprinting and immune refocusing in vaccine design. *Vaccine*, *26*(49), 6189–6199.

Trkola, A., Kuster, H., Rusert, P., Joos, B., Fischer, M., Leemann, C., et al. (2005). Delay of HIV-1 rebound after cessation of antiretroviral therapy through passive transfer of human neutralizing antibodies. *Nature Medicine*, *11*(6), 615–622.

van Gils, M. J., & Sanders, R. W. (2014). In vivo protection by broadly neutralizing HIV antibodies. *Trends in Microbiology*, *22*(10), 550–551.

Van Regenmortel, M. H. V. (2014). An outdated notion of antibody specificity is one of the major detrimental assumptions of the structure-based reverse vaccinology paradigm which prevented it from helping to develop an effective HIV-1 vaccine. *Frontiers in Immunology*, *5*, 1–8.

Venditto, V. J., Watson, D. S., Motion, M., Montefiori, D., & Szoka, F. C., Jr. (2013). Rational design of membrane proximal external region lipopeptides containing chemical modifications for HIV-1 vaccination. *Clinical and Vaccine Immunology*, *20*(1), 39–45.

Verkoczy, L., Chen, Y., Zhang, J., Bouton-Verville, H., Newman, A., Lockwood, B., et al. (2013). Induction of HIV-1 broad neutralizing antibodies in 2F5 knock-in mice: Selection against membrane proximal external region-associated autoreactivity limits T-dependent responses. *Journal of Immunology*, *191*(5), 2538–2550.

Verkoczy, L., Kelsoe, G., & Haynes, B. F. (2014). HIV-1 envelope gp41 broadly neutralizing antibodies: Hurdles for vaccine development. *PLoS Pathogens*, *10*(5), e1004073.

Vollmer, J., & Krieg, A. M. (2009). Immunotherapeutic applications of CpG oligodeoxynucleotide TLR9 agonists. *Advanced Drug Delivery Reviews*, *61*(3), 195–204.

Walker, B. D., & Burton, D. R. (2008). Toward an AIDS vaccine. *Science*, *320*(5877), 760–764.

Watkins, D. I. (2008). Highlights of the 15th conference on retroviruses and opportunistic infections. Basic HIV vaccine development. *Topics in HIV Medicine*, *16*(1), 7–8.

Watson, D. S., Endsley, A. N., & Huang, L. (2012). Design considerations for liposomal vaccines: Influence of formulation parameters on antibody and cell-mediated immune responses to liposome associated antigens. *Vaccine*, *30*(13), 2256–2272.

Watson, D. S., Platt, V. M., Cao, L., Venditto, V. J., & Szoka, F. C., Jr. (2011). Antibody response to polyhistidine-tagged peptide and protein antigens attached to liposomes via lipid-linked nitrilotriacetic acid in mice. *Clinical and Vaccine Immunology*, *18*(2), 289–297.

Watson, D. S., & Szoka, F. C., Jr. (2009). Role of lipid structure in the humoral immune response in mice to covalent lipid-peptides from the membrane proximal region of HIV-1 gp41. *Vaccine*, *27*(34), 4672–4683.

Wei, X., Decker, J. M., Wang, S., Hui, H., Kappes, J. C., Wu, X., et al. (2003). Antibody neutralization and escape by HIV-1. *Nature*, *422*(6929), 307–312.

West, A. P., Jr., Scharf, L., Scheid, J. F., Klein, F., Bjorkman, P. J., & Nussenzweig, M. C. (2014). Structural insights on the role of antibodies in HIV-1 vaccine and therapy. *Cell*, *156*(4), 633–648.

White, S. H., & Wimley, W. C. (1999). Membrane protein folding and stability: Physical principles. *Annual Review of Biophysics and Biomolecular Structure*, *28*, 319–365.

Xu, H., Song, L., Kim, M., Holmes, M. A., Kraft, Z., Sellhorn, G., et al. (2010). Interactions between lipids and human anti-HIV antibody 4E10 can be reduced without ablating neutralizing activity. *Journal of Virology*, *84*(2), 1076–1088.

Yang, G., Holl, T. M., Liu, Y., Li, Y., Lu, X., Nicely, N. I., et al. (2013). Identification of autoantigens recognized by the 2F5 and 4E10 broadly neutralizing HIV-1 antibodies. *The Journal of Experimental Medicine*, *210*(2), 241–256.

Zhang, J., Alam, S. M., Bouton-Verville, H., Chen, Y., Newman, A., Stewart, S., et al. (2014). Modulation of nonneutralizing HIV-1 gp41 responses by an MHC-restricted TH epitope overlapping those of membrane proximal external region broadly neutralizing antibodies. *Journal of Immunology*, *192*(4), 1693–1706.

Zhou, M., Kostoula, I., Brill, B., Panou, E., Sakarellos-Daitsiotis, M., & Dietrich, U. (2012). Prime boost vaccination approaches with different conjugates of a new HIV-1 gp41 epitope encompassing the membrane proximal external region induce neutralizing antibodies in mice. *Vaccine*, *30*(11), 1911–1916.

Zhu, Z., Qin, H. R., Chen, W., Zhao, Q., Shen, X., Schutte, R., et al. (2011). Cross-reactive HIV-1-neutralizing human monoclonal antibodies identified from a patient with 2 F5-like antibodies. *Journal of Virology*, *85*(21), 11401–11408.

CHAPTER THREE

Recombinant Lipoproteins as Novel Vaccines with Intrinsic Adjuvant

Pele Chong[1], Jui-Hsin Huang, Chih-Hsiang Leng, Shih-Jen Liu, Hsin-Wei Chen
Vaccine R&D Center, National Health Research Institutes, Miaoli, Taiwan
[1]Corresponding author: e-mail address: pelechong@nhri.org.tw

Contents

Abstract

A core platform technology for high production of recombinant lipoproteins with built-in immunostimulator for novel subunit vaccine development has been established. This platform technology has the following advantages: (1) easily convert antigen into lipidated recombinant protein using a fusion sequence containing lipobox and express high level (50–150 mg/L) in *Escherichia coli*; (2) a robust high-yield up- and downstream bioprocess for lipoprotein production is successfully developed to devoid endotoxin contamination; (3) the lipid moiety of recombinant lipoproteins, which is identical to that of bacterial lipoproteins is recognized as danger signals by the immune system (Toll-like receptor 2 agonist), so both innate and adaptive immune responses can be induced by lipoproteins; and (4) successfully demonstrate the feasibility and safety of

Advances in Protein Chemistry and Structural Biology, Volume 99
ISSN 1876-1623
http://dx.doi.org/10.1016/bs.apcsb.2015.03.003

this core platform technology in meningococcal group B subunit vaccine, dengue subunit vaccine, novel subunit vaccine against *Clostridium difficile*-associated diseases, and HPV-based immunotherapeutic vaccines in animal model studies.

1. INTRODUCTION

Subunit vaccines based on recombinant proteins or synthetic peptides are safer and cause fewer side effects than inactivated and live attenuated vaccines, but these well-defined vaccine candidates are usually poorly immunogenic and require appropriate adjuvant to be efficacious (Pardoll, 2002). The outer membrane lipoprotein, OspA, of *Borrelia burgdorferi*, which is a Braun's lipoproteins (BLP) (Erdile et al., 1993), has been licensed in the United States as human vaccine against Lyme disease. Animal immunized with the full-length OspA (lipidated form) was shown to be protected against *B. burgdorferi* challenge (Fikrig, Barthold, Kantor, & Flavell, 1990; Fikrig et al., 1992; Parker et al., 1995; Philipp et al., 1997). In contrast, vaccination with nonlipidated OspA was unable to elicit protective immunity against the pathogen (Bockenstedt, Fikrig, Barthold, Kantor, & Flavell, 1993; Fikrig et al., 1990, 1992; Johnson et al., 1995; Parker et al., 1995; Philipp et al., 1997). Although these results have clearly demonstrated the superior protective property of lipoproteins, the function roles of lipid moiety were not fully appreciated at that time. Now, the superior immunoprotective properties of lipoproteins are shown to trigger both innate and adaptive immunity (Infante-Duarte & Kamradt, 1997; Oftung, Wiker, Deggerdal, & Mustafa, 1997). Therefore to enhance the potency of recombinant protein vaccines, one can design and engineer immunogens as lipoproteins by covalently linking the antigen proteins with lipid moieties. The cysteine-linked diacyl lipid moiety of BLP is recognized as a danger signal by the immune system to induce antimicrobial activity and trigger host defense mechanisms through Toll-like receptor (TLR) stimulation (Brightbill et al., 1999; Thoma-Uszynski et al., 2001). Bacterial lipoproteins or BLPs are characterized by the presence of a lipobox motif, which is located in the C-terminal part of their leader peptide and contains a conserved cysteine residue, which is the target for *N*-acyl-*S*-diacylglyceryl-cysteinyl modification as shown in Fig. 1 (Hantke & Braun, 1973). Modification of the precursor protein is mediated by the consecutive activity of three enzymes: the phosphatidylglycerol–preprolipoprotein diacylglyceryl transferase responsible

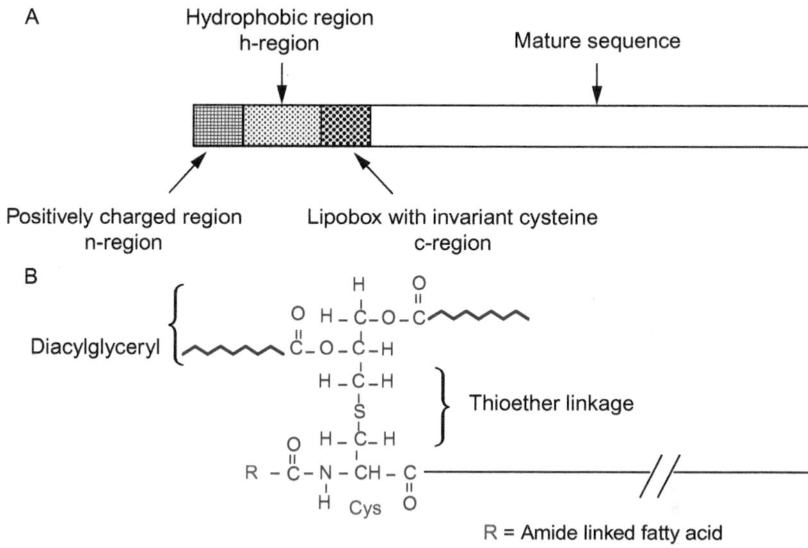

Figure 1 Schematic structure of recombinant lipoproteins. (A) The tripartite structure of lipoprotein signal sequence. The n-region has at least two positively charge residues; the h-region or the hydrophobic region is made up of 7–22 predominantly hydrophobic and uncharge residues; and the c-region which has the consensus [LVI]{ASTVI}{GAS}[C] sequence, is known as lipobox. (B) The structure of the lipid modification was elucidated by Hantke and Braun. The SH group of the N-terminal cysteine is modified with diacylglyceryl group attached through a thioether linkage, and the amino group was acylated with a fatty acid. *The figure is adapted from Chen et al. (2009) with permission from the publisher, Elsevier.*

for adding a diacylglycerol residue to the thiol group of the lipobox cysteine, the prolipoprotein signal peptidase/signal peptidase II which subsequently cleaves the lipidation signal sequence and the phospholipid–apolipoprotein *N*-acyltransferase which completes lipid modification (Hantke & Braun, 1973; Rezwan, Grau, Tschumi, & Sander, 2007). Although recombinant lipoproteins are recognized as potent vaccine candidates, they have not been fully used in subunit vaccine development because of poor lipoprotein expression. The posttranslational modifications (PTMs) and expression level of BLPs have been improved by fusing the target gene with selected signal peptide (SP) coding sequences (Cullen, Lo, Bulach, Cordwell, & Adler, 2003; De et al., 2000; Kamalakkannan, Murugan, Jagannadham, Nagaraj, & Sankaran, 2004; Legrain, Speck, & Jacobs, 1995), but the overexpression of lipoproteins in *Escherichia coli* frequently results in incomplete modification or the lack of lipidation (Hansson, Noppa, Nilsson, Stromqvist, & Bergstrom, 1995; Madurawe, Chase, Tsao, & Bentley,

2000; Shang, Summers, & Haake, 1996). Thus, large-scale production of lipoproteins remains a challenge.

2. ESTABLISHMENT OF HIGH EXPRESSION RECOMBINANT LIPOPROTEIN SYSTEM

Dumon-Seignovert, Cariot, and Vuillard (2004) had investigated the expression problems (low lipoprotein expression, no, or incomplete lipidated proteins) of lipoproteins in *E. coli*, and found several lipoproteins to be successfully expressed in strain BL21(DE3). However, the expression of recombinant lipoproteins in the mutant strains C41(DE3) and C43(DE3) was generally better than in BL21(DE3) due to high efficient transformation and lack of toxicity. It is not easy to express recombinant lipoproteins if the antigens contain no lipobox signal sequences. To express recombinant chimeric lipoproteins was finally succeeded when a potential *Neisseria meningitidis* lipoprotein vaccine candidate Ag473 was produced at high level in C43(DE3) cells (Chen et al., 2009; Hsu et al., 2008).

Using immunoproteomics technology, an outer membrane protein Ag473 (Genebank accession number: AY566590) was identified as a potential vaccine candidate by a protective mouse monoclonal antibody (MAb) generated against *N. meningitidis* (NM) group B bacteria (Hsu et al., 2008). Ag473 was found to contain the characteristic SP of BLPs, which has common structural features including a positively charged region (n-region), a hydrophobic region, (h-region), and a lipobox with an invariant cysteine (c-region) (Fig. 1). The c-region contains a conserved region of four amino acids, [LVI][ASTVI][GAS][C], known as lipobox, close to the cleavage site (Hayashi & Wu, 1990; von Heijne, 1989) and the invariant cysteinyl residue in the lipobox is the site for lipid modification (Hantke & Braun, 1973; Fig. 1B). The lipoprotein SP can be identified using the lipoprotein-prediction software (http://www.mrc-lmb.cam.ac.uk/genomes/dolop/analysis.shtml).

The *Ag473* gene was engineered with a His-tag at the C-terminal and cloned into pET-22b under the control of the T7 promoter. Three *E. coli* strains, JM109(DE3), BL21(DE3), and C43(DE3) were tested for recombinant Ag473 (rAg473) expression after isopropyl β-D-thiogalactoside (IPTG) by SDS-PAGE analysis. No obvious induction was observed with JM109(DE3) cells. However, rAg473 was induced in BL21(DE3) cells and even at a higher level in C43(DE3) cells (Fig. 2A). rAg473 was initially purified from *E. coli* lysate supernatants by immobilized

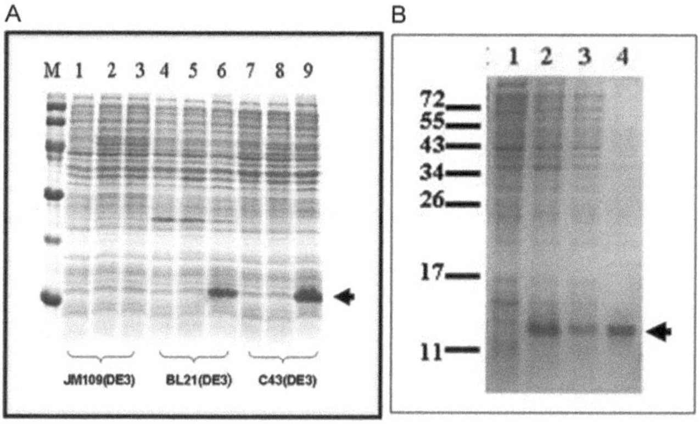

Figure 2 Expression and purification of rAg473. (A) The recombinant Ag473 (rAg473) was expressed in three different *E. coli* strains, JM109(DE3), BL21(DE3), and C43(DE3). Lanes 1, 4, and 7, untransformed plasmids; lanes 2, 5, and 8, protein expression in the absence of IPTG induction; lanes 3, 6, and 9, protein expression after IPTG induction. (B) Immunoblot monitoring of the purification process. rAg473 was detected using anti-$(His)_6$ antibodies. The arrow indicates the electrophoretic position of rAg473 in the gels. *The figure is adapted from Chen et al. (2009) with permission from the publisher, Elsevier.*

metal affinity chromatography in the absence of detergent (Fig. 2B). Routinely, 4–5 mg of purified rAg473 protein were obtained from 250 mL of shake flask culture (harvest $OD_{600}=0.7$) corresponding to a 20% yield. Subsequently, the purified rAg473 was subjected to N-terminal amino acid sequencing using Edman degradation. The results showed that rAg473 expressed in BL21(DE3) started with MKKL that indicated this rAg473 was not fully lipidated. In contrast, the N-terminal sequence of rAg473 expressed in C43(DE3) could not be obtained that suggested the N-terminus was blocked. Since the N-terminal end of bacterial lipoproteins is blocked as a result of PTMs (Fig. 1) that is consistent with rAg473 expressed as a BLP. The purified rAg473 was further characterized using mass spectrometry to demonstrate to have the correct lipid modification and immunological properties (Kwok et al., 2011; Sung et al., 2010).

3. THE LIPID STRUCTURE OF LIPOPROTEIN

The structural analysis of PTMs of lipoproteins is difficult due to the hydrophobic properties of their fatty acid moieties. Direct sample infusion ESI-MS of purified rAg473 was performed on a Q-TOF instrument (Waters Q-TOF). After processing the data using the maximum entropy deconvolution algorithm MaxEnt1 (Waters, Milford, MA, USA), three major peaks with masses of 12976, 12990.5, and 13004 Da, respectively, were identified (Fig. 3A). Interestingly, two contiguous peaks differ in mass by 14 atomic mass unit (amu) (Fig. 3A). The existence of these three peaks was likely the signature for PTMs of rAg473 expressed in C43 (DE3) cells. Hence, the incremental differences of 14 amu between peaks were caused by different posttranslational lipid modifications and resulting

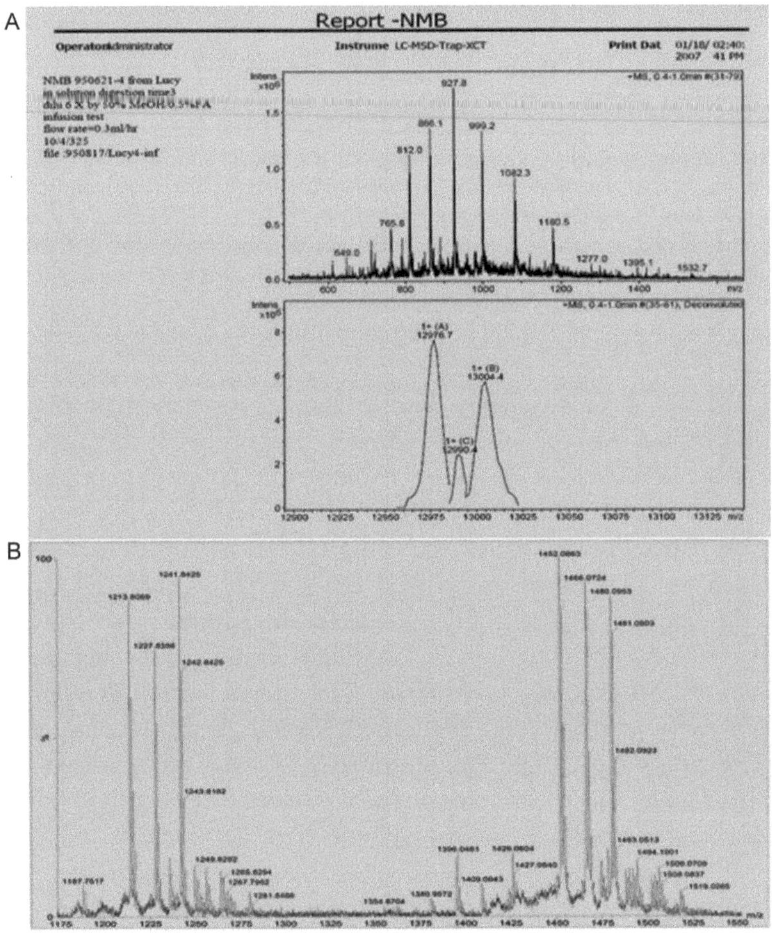

Figure 3 LC/MS analysis of rAg473 and MALDI-TOF analysis of tryptic digested lipopeptide fragments. (A) The purified rAg473 was analyzed by LC/MS and revealed the presence of three major posttranslational modifications of rAg473, with each modified form differing mass by 14 amu (12976.7, 12990.7, and 13004.7, respectively). (B) The N-terminal tryptic digested fragments was analyzed by MALDI-TOF and revealed two groups I and II. *The figure is adapted from Sung et al. (2010) with permission from the publisher, Elsevier.*

in heterogeneous rAg473 species. To test this hypothesis, the purified rAg473 was digested with trypsin and the tryptic fragments were analyzed using MALDI-TOF. Two groups of ion mass peaks exhibit the PTMs signature from the mass spectra analysis of tryptic digested rAg473 (Fig. 3B). Group I contains three peaks with *m/z* values of 1451.9, 1465.9, and 1479.9 (Fig. 3B), whereas group II consists of three peaks with value of *m/z* 1936.2, 1950.2, and 1964.2, respectively (Fig. 3B). The mass differences between these peaks are 14 amu and the pattern of each group is exactly identical to that of rAg473 (Fig. 3A). The N-terminal fragments of rAg473 and the putative sequences of group I and II lipopeptides were identified that group I fragments correspond to the lipid-modified CSQEAK sequence whereas in group II fragments correspond to lipidated CSQEAKQEVK peptides based on the direct sequencing of *y*-series fragments using collision-induced dissociation (CID) and MS^n analysis (Kwok et al., 2011; Sung et al., 2010).

The identity of lipid moiety of the tryptic fragments derived from rAg473 were revealed using GC/MS and MS^n analysis (Kwok et al., 2011; Sung et al., 2010). Figure 4 shows the molecular structure of CID

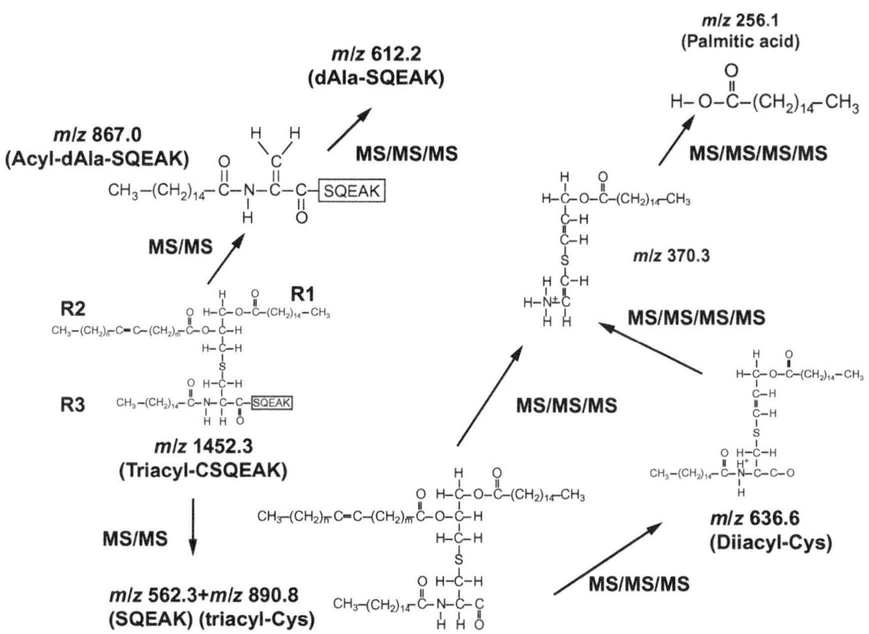

Figure 4 Molecular structure of chemical-induced dissociation (CID) fragmentation of triacyl peptide and its corresponding MS ion detected in MS^n analysis.

fragmentation of triacyl lipopeptide and its corresponding MS ion detected in MS^n analysis. Compounds (A) to (G) demonstrated the results of each fragmentation step and major ion mass generated in the process. The bacterial lipid structure was identified and reported the first time at the molecular level (Kwok et al., 2011; Sung et al., 2010). R1, R2, and R3 represent the fatty acids position of sn1, sn2, and N-terminal amine on lipid moiety (Kwok et al., 2011). Multiple-stage fragmentation analyses of the di- and triacyl lipopeptides using both the positive and negative ion modes enabled to identify the putative structure of the *N*-acyl-*S*-diacylglycerylcysteine containing an amide bond to palmitic acid at the N-terminal cysteine (R3 position) and a palmitic acid at R1 position. Importantly, an unsaturated fatty acid of either hexadecenoic acid, cyclopropaneoctanoic acid, oleic acid, or nonadecenoic acid at R2 position of diacylglycerol residue through ester bonding has been identified (Kwok et al., 2011; Sung et al., 2010). The result clearly showed that the lipid moieties of rAg473 contain an unsaturated bond at R2 which is different from that of synthetic tripalmitoylated lipopeptide, Pam3Cys.

4. RECOMBINANT LIPOPROTEIN WITH INTRINSIC ADJUVANT PROPERTIES

Like most characterized lipoproteins, the lipid moiety of the Ag473 protein is required for the self-adjuvant activity and this property was attributed to its ability to activate and promote bone marrow-derived dendritic cells (BMDCs) maturation (Reise Sousa, 2006). According to the principle of vaccine, activation of antigen-presenting cells (APCs) is required for inducing the protective immune responses. Dendritic cells (DCs) are professional APCs, which are critical in the initiation of immune responses during infection or vaccination (Joffre, Nolte, Spörri, & Reise Sousa, 2009; Lopez-Bravo & Ardavin, 2008; Medzhitov, 2007). DCs capture antigens and then migrate to secondary lymphoid organs where they present processed antigens to stimulate antigen-specific T cells. DCs become mature after stimulating by typical microbial molecules or pathogen-associated molecular patterns (PAMPs) and enhance the ability of activating T cells (Chu et al., 2012; Ishii, Koyama, Nakagawa, Coban, & Akira, 2008; Joffre et al., 2009; Lopez-Bravo & Ardavin, 2008; Medzhitov, 2007; Takeuchi & Akira, 2010). TLRs are the most important pattern recognition receptors to detect PAMPs in DCs (Ishii et al., 2008; Joffre et al., 2009; Lopez-Bravo & Ardavin, 2008; Medzhitov, 2007; Reise Sousa, 2006;

Takeuchi & Akira, 2010). TLR ligand induces the release of both cytokines and chemokines and the upregulation of MHC class II and costimulatory molecule expression (Chu et al., 2012; Ishii et al., 2008; Medzhitov, 2007; Takeuchi & Akira, 2010). The immunomodulating activities of lipoprotein are mediated primarily by TLR2 in cooperating with TLR1 and/or TLR6. The detections of IL-12 in rAg473 treated BMDCs and IFN-γ in the recall assays, indicating that rAg473 has an ability to skew the immune response toward Th1 response. This is in consistence with the *in vivo* antibody response which showed that substantial levels of IgG2 were detected in the rAg473 immune serum in addition to the IgG1. The ratios of IgG1/IgG2 in the rAg473 and nonlipidated Ag473 immune sera were approximately 0.75 and 2, respectively. Balb/c mice immunized with 10 μg of rAg473 produced $>10^6$ anti-Ag473 IgG titer and protected against a lethal meningococcal infection (Table 1). The current data indicate that rAg473 has an intrinsic adjuvant with capability to induce both humoral and cell-mediated protective immune responses. In addition, the responses of $TLR2^{-/-}$ and wild-type DC to rAg473 stimulation were comparable but dramatically reduced in TLR4-defective DCs, suggesting that rAg473 could be a TLR4 agonist. This is not unprecedented since lipoproteins BmpA of *B. burgdorferi* and Omp16 of *Brucella abortus* were reported to signal via TLR4 and such properties were attributed to the protein portion (Pasquevich et al., 2010; Yang et al., 2008). The observation that nonlipidated Ag473,

Table 1 Anti-Ag473 IgG Titers of Antisera from Groups of 6 Mice Immunized Three Times with rAg473 or Nonlipidated Ag473 Formulated with or Without CFA/IFA

Mice Immunized with		Anti-Ag473 IgG Titers[a]	The Protection Rate (%) after Challenged with[b]	
			100 cfu	10 cfu
rAg473	3 μg	1.1×10^5	0	30
	10 μg	5.2×10^6	40	100
	10 μg + CFA/IFA	8.1×10^6	30	100
Nonlipidated Ag473	10 μg	7.7×10^4	0	20
	10 μg + CFA/IFA	3.7×10^6	40	70

[a]The titers were obtained with sera pooled from six mice.
[b]Groups of 6 mice were immunized three times with different Ag473 formulated with or without CFA/IFA, then challenged with different amount of live meningococcal group B bacteria. The survival rate in each group was reported post 36 h challenge.

heat-treated, or proteinase K-treated rAg473, failed to induce the production of TNF-α by DCs, suggesting that rAg473 signaling via TLR is dependent on certain conformation formed by both the lipid moiety and the protein moiety. Similar to LPS molecules, rAg473 may assume various supramolecular structures which may affect the TLR dependence.

5. TO ENGINEER CHIMERIC LIPOPROTEINS WITH INTRINSIC ADJUVANT FOR NOVEL VACCINE DEVELOPMENT

As rAg473 is overproduced in the C43(DE3) expression system, it is of interest to explore whether this recombinant lipoprotein expression system could be applied to engineer and convert nonlipidated antigens into lipidated immunogens. To explore whether it is possible to engineer potent lipoimmunogens, as a model study the N-terminal residues of Ag473 is fused with the domain 3 of the dengue virus envelope protein (E3), which contains the protective epitopes and receptor recognition site of the envelope (E) of dengue viruses. The fusion of E3 with the Ag 473 SP did not result in the production of a recombinant lipoimmunogen. To determine if other lipid SPs could yield better results, E3 was fused with other SPs identified from *E. coli* proteins such as TraK lipoprotein precursor, the peptidoglycan-associated lipoprotein precursor, the acriflavine-resistance protein E precursor, the membrane-bound lytic murein transglycosylase D precursor, and the flagellar L-ring protein precursor. After transformation into *E. coli* C43(DE3), none of these constructs expressed noticeable amounts of recombinant lipoimmunogen after IPTG induction (Fig. 5A; Chen et al., 2009; Hayashi & Wu, 1990; von Heijne, 1989). These results suggested that additional sequences beyond the signal lipobox were necessary to produce viral lipoimmunogens in scalable quantities. Therefore, the E3 was fused to the C-terminus of three different N-terminal sequences of Ag473 varying in length. According to its primary sequence, the Ag473 protein can be divided into four domains: SP, domain 1 (d1), domain 2 (d2), and domain 3 (d3), respectively. Four constructs were engineered in which E3 was fused with SP, SP/d1, SP/d1/d2, and SP/d1/d2/d3 resulting in SPE3, D1E3, D2E3, and D3E3 fusion proteins, respectively (Fig. 5A). SPE3 was not expressed after IPTG induction. In contrast, the expression of D1E3, D2E3, and D3E3 was evident (Fig. 5B). Moreover, the N-terminal sequences for recombinant D1E3, D2E3, and D3E3 could not be obtained by Edman degradation, these recombinant antigens were digested with

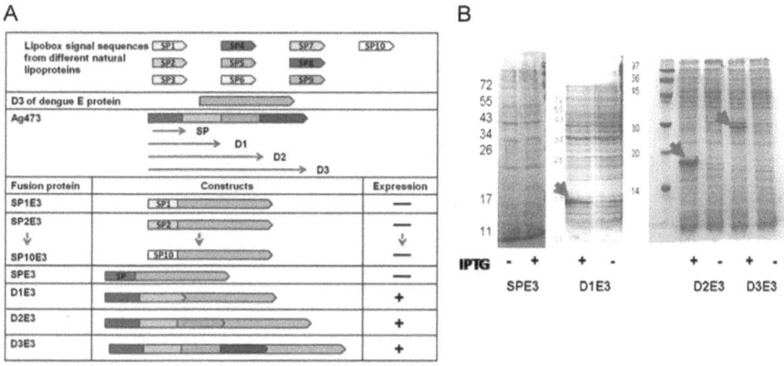

Figure 5 Construction and expression of fusion E3 antigens with different lipobox sequences. (A) Depicts the plasmid construction for expression of recombinant fusion protein comprising the signal peptide (SP1–SP10) from different natural lipoprotein, signal peptide (SP), domain 1 (D1), domain 2 (D2), or domain 3 (D3) from Ag473, and the dengue antigen E3 sequence, designated SPE3, D1E3, D2E3, and D3E3, respectively. (B) The IPTG induction of fusion protein expression in *E. coli* strain C43(DE3). *The figure is adapted from Chen et al. (2009) with permission from the publisher, Elsevier.*

trypsin and their tryptic fragments were analyzed using MALDI-TOF. The mass determinations were identical to those obtained from rAg473, so these recombinant antigens exhibit the characteristics of lipoprotein and the post-translational lipid modifications are identical to those of rAg473.

6. RECOMBINANT CHIMERIC LIPOPROTEINS CAN ACTIVATE BMDCs

BMDCs could be activated by recombinant chimeric lipoprotein and the lipoprotein-induced cytokines from BMDCs are mediated by TLR2 (Ishii et al., 2008; Takeuchi & Akira, 2010). Figure 6 shows that rlipo-D1E3 induced the activation of spleen cells and BMDCs in wild-type and TLR4-deficient mice, but not in TLR2$^{-/-}$ mice (Chen et al., 2009). These results are different from those obtained from rAg473 which can induced DC maturation in TLR2$^{-/-}$ mice, and support that rAg473 may assume various supramolecular structures which may affect the TLR dependence. Further studies (Chen et al., 2009; Leng et al., 2010) indicated that rlipo-D1E3 induced significant production of TNF-α, IL-6, and IL-12p40 from the BMDCs of wild-type and TLR4-deficient mice but not from the BMDCs of TLR2$^{-/-}$ mice ($p < 0.01$). rlipo-D1E3 immunized wild-type and TLR4-deficient mice produced equivalent IgG titers of anti-rE3 antibodies, but anti-rE3 antibody titers were decreased when

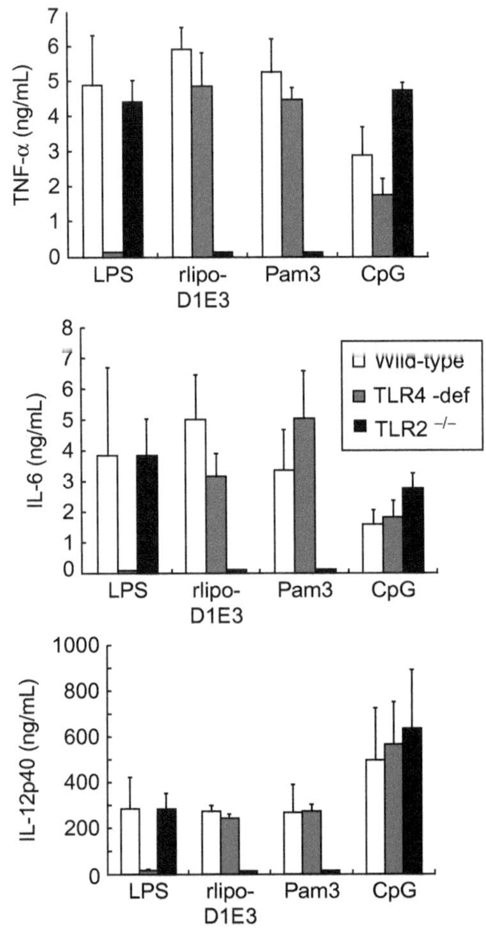

Figure 6 The effector responses of BMDCs after stimulation with rlipo-D1E3. BMDCs from wild-type, TLR4-deficient, and $TLR2^{-/-}$ mice were cultured in medium alone or in medium supplemented with LPS (0.01 μg/mL), rlipo-D1E3 (100 nM), Pam3 (100 nM), or CpG (10 μg/mL). After incubation for 24 h, the supernatants were harvested and analyzed for TNF-α, IL-6, and IL12p40 by ELISA. *The figure is adapted from Leng et al. (2010) with permission from the publisher, Elsevier.*

$TLR2^{-/-}$ mice were immunized. rlipo-D1E3 and Pam3 could stimulate BMDCs from $TLR1^{-/-}$ or $TLR6^{-/-}$ mice to secrete TNF-α, but MALP2 could not stimulate BMDCs from $TLR6^{-/-}$ mice to secrete TNF-α (Leng et al., 2010). These data further show that rlipo-D1E3 does not act through TLR1- or TLR6-dependent TLR2 activation. As rAg473 and rlipo-D1E3 containing an unsaturated lipid at R2 which is different from that of Pam3, were found to induce higher levels of IL-23, IL-27, and MIP-1α secretion than Pam3. These results suggest that recombinant lipoproteins (rAg473 and rlipo-D1E3) and Pam3 activate BMDCs through the same receptor (TLR2) and a similar MAPK signaling pathway, but they may induce different gene expression levels to regulate immune responses (Leng et al., 2010).

7. A ROBUSTNESS UP- AND DOWNSTREAM PROCESSES FOR LIPOPROTEIN PRODUCTION

Because without optimizing the growth conditions some lipoproteins were not lipidated, a robustness upstream process for lipoprotein expression was further investigated. For high-level expression of mature lipoproteins in the C43(DE3) strain, M9 medium was found to be better than M63 culture medium and other rich media (Tseng & Leng, 2012). Furthermore, the influence of media factors (including nitrogen and carbon sources, phosphate, ferrous ions, calcium, magnesium, and pH) was studied, analyzed, and found to have great impact on the levels of lipoprotein expression. After testing in many chimeric lipoprotein systems (see below), M9 medium is validated to be the cost-effective and best upstream culture medium for high-level expression of lipoproteins in the C43(DE3) strain.

In order to have high lipoprotein expression, the DNA coding sequence of the target antigen should use codon usage optimized, chemically synthesized, and inserted into pET-22b vector for expressing in *E. coli* strain C43 (DE3). After IPTG induction, in general, highly purified recombinant lipoproteins (>95% purity) are obtained by a single-step purification using Ni-affinity chromatography in the presence of 200–500 mM NaCl and 0.1% of TritonX-100. The majority of LPS, high salt and TritonX-100, are normally removed during the washing step with low concentration of imidazole buffer. Recombinant lipoprotein was eluted using PBS buffer containing 100–500 mM imidazole and dialyzed in a 30-kDa cut-off dialysis bag against phosphate-buffered saline (PBS) containing 10% glycerol. The purity of recombinant lipoproteins is confirmed by SDS-PAGE and the western blot analysis using either the antigen-specific MAb or the anti-His-tag MAb. Trace amounts of degradation fragments of lipoproteins were also detected using antigen-specific antibody. These degradation products are likely the result of proteolytic digestion during the purification process. In any event, at least 5–10 mg of highly enriched recombinant lipoproteins could be easily obtained from 1 L of bacteria culture. *E. coli* LPS were further successfully removed by passing the lipoprotein preparation through an

E membrane; and the residual LPS in the purified lipoprotein would be below 30 EU/mL based on the Limulus assay. Although a simple and rapid method for producing lipoproteins with high purity was successfully developed, some recombinant lipoproteins were found to be unstable and easily lost its biological function during the freeze–thaw process. We found that the best condition for preserving lipoprotein integrity was to store the protein at 1 mg/mL in PBS containing 10% (v/v) of glycerol at −80 °C.

8. RECOMBINANT LIPOPROTEIN-BASED SUBUNIT VACCINE INDUCES HIGH TITERS OF NEUTRALIZING ANTIBODY AND LONGER MEMORY IMMUNE RESPONSES

To validate the lipoprotein technology, a novel lipoprotein-based dengue subunit vaccine (rlipo-D1E3) as described above was rationally designed to contain the consensus amino acid sequences of domain 3 derived from the E protein of four serotypes of dengue viruses (Chen et al., 2009). As shown in Figs. 5 and 6, rlipo-D1E3 is capable of activating APCs and enhancing cellular and humoral immune responses. Further studies demonstrated that a single dose of rlipo-D1E3 immunization in mice without extra adjuvant formulation is sufficient to elicit neutralizing antibodies against all four serotypes of dengue viruses (Chen et al., 2009; Chiang et al., 2011). In addition, strong memory responses were elicited in mice immunized with a single dose of rlipo-D1E3. Quick, anamnestic-neutralizing antibody responses to a live dengue virus challenge were elicited at week 28 post-immunization. These results demonstrate the promising possibility of a future successful tetravalent vaccine against dengue viral infections that utilizes one-dose vaccination with rlipo-D1E3.

Recombinant lipoprotein technology is also applied to subunit vaccine development against *Clostridium difficile* infections (CDI) that is an opportunistically nosocomial infections in hospitalized patients. Clostridial exotoxins A and B (TcdA and TcdB) specifically bind to unknown glycoprotein(s) in the host intestine, disrupt the intestinal barrier, and lead to acute inflammation and diarrhea. The C-terminal receptor binding domain (>900 amino acids) of TcdA (A-rRBD) has been shown to elicit antibody responses in mice and rabbits that neutralize TcdA toxicity in the Vero cell cytotoxicity assay, but not effectively protect hamsters against a lethal dose of *C. difficile* spores in the hamster challenge model. To develop an effective recombinant subunit vaccine against CDI, A-rRBD was lipidated (rlipoA-RBD) as a rational design to contain an intrinsic adjuvant, TLR2 agonist and expressed in *E. coli* strain C43(DE3). The purified rlipoA-RBD was characterized immunologically and found to have the following properties: (a) mice, hamsters, and rabbits vaccinated with 3 μg of rlipoA-RBD produced strong antibody responses that could neutralize TcdA toxicity in the Vero cell cytotoxicity assay and the neutralization titer was comparable to those obtained from antisera immunized either with 10 μg of TcdA toxoid or 30 μg of rA-rRBD; (b) rlipoA-RBD elicited immune responses and protect mice from TcdA challenge, but insignificant protection (10–20%) against *C. difficile* spores challenge in the hamsters model; (c) only rlipoA-RBD formulated with TcdB RBD consistently confers protection (90–100%) in the hamsters challenge model; and (d) rlipo-A-RBD was found to be 10-fold more potent than A-rRBD as adjuvant to enhancing immune responses against poor antigen such as ovalbumin (Fig. 7). These results indicate that rlipoA-RBD formulated with B-rRBD could be an excellent vaccine candidate for preclinical studies and future clinical trials.

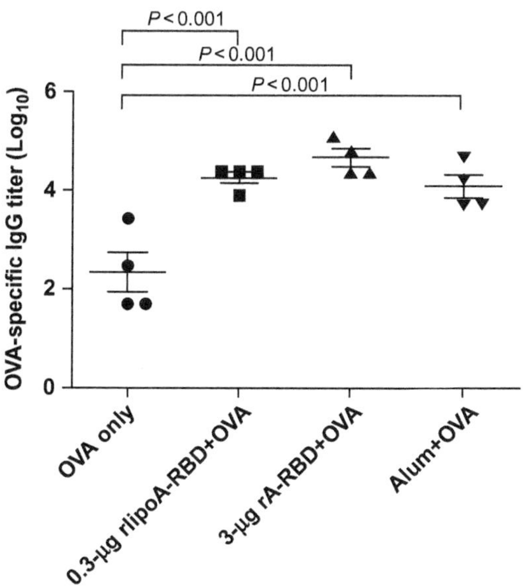

Figure 7 Adjuvant effect of rlipoA-RBD as demonstrated by the enhancement of anti-OVA IgG response by coadministration of rlipoA-RBD and OVA. BALB/c mice were immunized with 2 × 2 μg of OVA formulated with either 0.3 μg of rlipoA-RBD or 3 μg of rA-RBD or formulated with alum as the positive control. Anti-OVA IgG titer was determined by OVA-specific ELISA.

9. RECOMBINANT LIPIDATED HUMAN PAPILLOMAVIRUS E7 INDUCES CELLULAR IMMUNE RESPONSES AND PROTECTIVE IMMUNITY AGAINST CERVICAL CANCER IN THE TC-1 MOUSE MODEL

It is well documented that protein-based immunogens alone elicit little or no cytotoxic T-lymphocyte (CTL) responses. Recombinant proteins have been difficult to exploit for immunotherapeutic purposes. Recombinant lipoprotein technology was further applied to the development of therapeutic vaccine (Huang et al., 2012). The oncoprotein E7 of human papillomavirus (HPV) is one of the targets for developing immunotherapeutic vaccines against HPV-associated tumors. We recently reported (Huang et al., 2012) that a recombinant lipoprotein consisting of inactive E7 (E7m) biologically linked to a bacterial lipid moiety (rlipo-E7m) induces the maturation of mouse BMDCs through TLR2, skews the immune responses toward the Th1 responses and induces E7-specific CTL responses. As shown in the study (Huang et al., 2012), the ability of rlipo-E7m to provide protection against a TC-1 tumor cell challenge in an animal model. Mice prophylactically immunized with two 10-μg doses of rlipo-E7m were found to be free of TC-1 tumor growth (Fig. 8). Experiments in a therapeutic immunization model showed that the tumor volume in mice receiving a single dose of rlipo-E7m was less than 0.01 cm^3 on day 40, whereas the tumor volume in mice treated with rE7m was 2.28 ± 1.21 cm^3 (Fig. 8A). The tumor volume of the negative control group was over 3 cm^3. In addition, we demonstrated that the CD8+ T cells play a major role in antitumor immunity since the immunotherapeutic protective effect of rlipo-E7m was totally abolished when the immunized mice treated with a MAb specific against CD8+ T cells (Fig. 8B). These results demonstrate that rlipo-E7m could be a promising candidate for treating HPV-associated tumors.

10. CONCLUSION

We have demonstrated that nonlipidated antigens with molecular size ranging from 15 kDa (rlipo-D1E3) to 100 kDa (rlipoA-RBD) can be efficiency converted into lipidated antigens and produced in a cost-effective up- and downstream bioprocess in *E. coli* strain C43(DE3). These lipoproteins are capable of stimulating the BMDCs through TLR2, eliciting strong

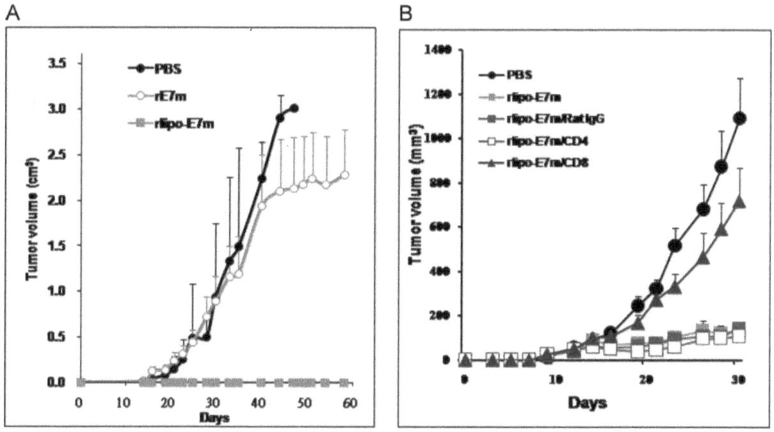

Figure 8 Immunization with rlipo-E7m induces a strong antitumor effect. (A) Mice were immunized twice by subcutaneous injection of rE7m/rlipo-E7m (10 μg) or PBS at 1-week intervals. Seven days after the final immunization, mice were subcutaneously injected with 26105 TC-1 cells in a total volume of 200 mL. (B) Three groups (rlipo-E7m/CD4, rlipo-E7m/CD8, and rlipo-E7m/Rat IgG) of mice were injected of anti-CD4, anti-CD8, and control antibodies, respectively, 1 day before the injection of TC-1 cells. These groups and two additional groups (rlipo-E7m and PBS) were injected once with rlipo-E7m (10 μg/mouse) or PBS 7 days after inoculated with TC-1 cells. The tumor volume was shown as length × width × width/2 (mm^3). *The figure is adapted from Huang et al. (2012) with permission from the publisher, PLoS organization.*

and long memory immune response, and inducing antigen-specific CTL responses. In prophylactic vaccine, the lipidated immunogen alone was able to elicit strong functional virus-neutralizing antibody responses better than those obtained from nonlipidated antigen formulated with alum adjuvant. In therapeutic vaccine, the lipidated immunogen could generate tumor-protective immunity in a mouse tumor model. This strategy certainly provides a method for developing successful prophylactic and therapeutic vaccines using lipoprotein-based immunogens that are safe and cost-effective vaccines for human use. All the results described above demonstrate the merit of lipoproteins with intrinsic adjuvant properties for novel vaccine development.

ACKNOWLEDGMENT

This work was supported by the grants from the Ministry of Science and Technology, and National Health Research Institutes, Taiwan.

REFERENCES

Bockenstedt, L. K., Fikrig, E., Barthold, S. W., Kantor, F. S., & Flavell, R. A. (1993). Inability of truncated recombinant OspA proteins to elicit protective immunity to *Borrelia burgdorferi* in mice. *Journal of Immunology, 151*, 900–906.

Brightbill, H. D., Libraty, D. H., Krutzik, S. R., Yang, R. B., Belisle, J. T., Bleharski, J. R., et al. (1999). Host defense mechanisms triggered by microbial lipoproteins through toll-like receptors. *Science, 285*, 732–736.

Chen, H. W., Liu, S. J., Liu, H. H., Kwok, Y., Lin, C. L., Lin, L. H., et al. (2009). A novel technology for the production of a heterologous lipoprotein immunogen in high yield has implications for the field of vaccine design. *Vaccine, 27*, 1400–1409.

Chiang, C. Y., Liu, S. J., Tsai, J. P., Li, Y. S., Chen, M. Y., Liu, H. H., et al. (2011). A novel single-dose dengue subunit vaccine induces memory immune responses. *PloS One, 6*, e23319.

Chu, C. L., Yu, Y. L., Kung, Y. C., Liao, P. Y., Tseng, Y. T., Liu, K. J., et al. (2012). The immunomodulatory activity of meningococcal lipoprotein Ag473 depends on the conformation made up of the lipid and protein moieties. *PloS One*, 7(7), e40873.

Cullen, P. A., Lo, M., Bulach, D. M., Cordwell, S. J., & Adler, B. (2003). Construction and evaluation of a plasmid vector for the expression of recombinant lipoproteins in *Escherichia coli*. *Plasmid, 49*, 18–29.

De, B. K., Sampson, J. S., Ades, E. W., Huebner, R. C., Jue, D. L., Johnson, S. E., et al. (2000). Purification and characterization of *Streptococcus pneumoniae* palmitoylated pneumococcal surface adhesin A expressed in *Escherichia coli*. *Vaccine, 18*, 1811–1821.

Dumon-Seignovert, L., Cariot, G., & Vuillard, L. (2004). The toxicity of recombinant proteins in *Escherichia coli*: A comparison of overexpression in BL21(DE3), C41(DE3), and C43(DE3). *Protein Expression and Purification, 37*, 203–206.

Erdile, L. F., Brandt, M. A., Warakomski, D. J., Westrack, G. J., Sadziene, A., Barbour, A. G., et al. (1993). Role of attached lipid in immunogenicity of *Borrelia burgdorferi* OspA. *Infection and Immunity, 61*, 81–90.

Fikrig, E., Barthold, S. W., Kantor, F. S., & Flavell, R. A. (1990). Protection of mice against the Lyme disease agent by immunizing with recombinant OspA. *Science, 250*, 553–556.

Fikrig, E., Telford, S. R., 3rd, Barthold, S. W., Kantor, F. S., Spielman, A., & Flavell, R. A. (1992). Elimination of Borrelia burgdorferi from vector ticks feeding on OspA-immunized mice. *Proceedings of the National Academy of Sciences of the United States of America, 89*, 5418–5421.

Hansson, L., Noppa, L., Nilsson, A. K., Stromqvist, M., & Bergstrom, S. (1995). Expression of truncated and full-length forms of the Lyme disease *Borrelia* outer surface protein A in *Escherichia coli*. *Protein Expression and Purification, 6*, 15–24.

Hantke, K., & Braun, V. (1973). Covalent binding of lipid to protein. Diglyceride and amide-linked fatty acid at the N-terminal end of the murein-lipoprotein of the Escherichia coli outer membrane. *European Journal of Biochemistry, 34*, 284–296.

Hayashi, S., & Wu, H. C. (1990). Lipoproteins in bacteria. *Journal of Bioenergetics and Biomembranes, 22*, 451–471.

Hsu, C. A., Lin, W. R., Li, J. C., Liu, Y. L., Tseng, Y. T., Chang, C. M., et al. (2008). Immunoproteomic identification of the hypothetical protein NMB1468 as a novel lipoprotein ubiquitous in *Neisseria meningitidis* with vaccine potential. *Proteomics, 8*, 2115–2125.

Huang, C. Y., Chen, H. W., Shen, K. Y., Chang, L. S., Yeh, Y. C., Chen, I. H., et al. (2012). Recombinant lipidated HPV E7 induces a Th-1-biased immune response and protective immunity against cervical cancer in a mouse model. *PloS One*, 7, e40970.

Infante-Duarte, C., & Kamradt, T. (1997). Lipopeptides of *Borrelia burgdorferi* outer surface proteins induce Th1 phenotype development in alphabeta T-cell receptor transgenic mice. *Infection and Immunity, 65*, 4094–4099.

Ishii, K. J., Koyama, S., Nakagawa, A., Coban, C., & Akira, S. (2008). Host innate immune receptors and beyond: Making sense of microbial infections. *Cell Host & Microbe*, *3*, 352–363.

Joffre, O., Nolte, M. A., Spörri, R., & Reise Sousa, C. (2009). Inflammatory signals in dendritic cell activation and the induction of adaptive immunity. *Immunological Reviews*, *227*, 234–247.

Johnson, B. J., Sviat, S. L., Happ, C. M., Dunn, J. J., Frantz, J. C., Mayer, L. W., et al. (1995). Incomplete protection of hamsters vaccinated with unlipidated OspA from *Borrelia burgdorferi* infection is associated with low levels of antibody to an epitope defined by mAb LA-2. *Vaccine*, *13*, 1086–1094.

Kamalakkannan, S., Murugan, V., Jagannadham, M. V., Nagaraj, R., & Sankaran, K. (2004). Bacterial lipid modification of proteins for novel protein engineering applications. *Protein Engineering, Design & Selection*, *17*, 721–729.

Kwok, Y., Sung, W. C., Lin, A. L., Liu, H. H., Chou, F. A., Hsieh, S. S., et al. (2011). Rapid isolation and characterization of bacterial lipopeptides using liquid chromatography and mass spectrometry analysis. *Proteomics*, *11*, 2620–2627.

Legrain, M., Speck, D., & Jacobs, E. (1995). Production of lipidated meningococcal transferrin binding protein 2 in *Escherichia coli*. *Protein Expression and Purification*, *6*, 570–578.

Leng, C. H., Chen, H. W., Chang, L. S., Liu, H. H., Liu, H. Y., Sher, Y. P., et al. (2010). A recombinant lipoprotein containing an unsaturated fatty acid activates NF-kappaB through the TLR2 signaling pathway and induces a differential gene profile from a synthetic lipopeptide. *Molecular Immunology*, *47*, 2015–2021.

Lopez-Bravo, M., & Ardavin, C. (2008). In vivo induction of immune responses to pathogens by conventional dendritic cells. *Immunity*, *29*, 343–351.

Madurawe, R. D., Chase, T. E., Tsao, E. I., & Bentley, W. E. (2000). A recombinant lipoprotein antigen against Lyme disease expressed in *E. coli*: Fermentor operating strategies for improved yield. *Biotechnology Progress*, *16*, 571–576.

Medzhitov, R. (2007). Recognition of microorganisms and activation of the immune response. *Nature*, *449*, 819–826.

Oftung, F., Wiker, H. G., Deggerdal, A., & Mustafa, A. S. (1997). A novel mycobacterial antigen relevant to cellular immunity belongs to a family of secreted lipoproteins. *Scandinavian Journal of Immunology*, *46*, 445–451.

Pardoll, D. M. (2002). Spinning molecular immunology into successful immunotherapy. *Nature Reviews. Immunology*, *2*, 227–238.

Parker, T. S., Levine, D. M., Chang, J. C., Laxer, J., Coffin, C. C., & Rubin, A. L. (1995). Reconstituted high-density lipoprotein neutralizes gram-negative bacterial lipopolysaccharides in human whole blood. *Infection and Immunity*, *63*, 253–258.

Pasquevich, K. A., García Samartino, C., Coria, L. M., Estein, S. M., Zwerdling, A., Ibañez, A. E., et al. (2010). The protein moiety of *Brucella abortus* outer membrane protein 16 is a new bacterial pathogen-associated molecular pattern that activates dendritic cells in vivo, induces a Th1 immune response, and is a promising self-adjuvanting vaccine against systemic and oral acquired brucellosis. *Journal of Immunology*, *184*, 5200–5212.

Philipp, M. T., Lobet, Y., Bohm, R. P., Jr., Roberts, E. D., Dennis, V. A., Gu, Y., et al. (1997). The outer surface protein A (OspA) vaccine against Lyme disease: Efficacy in the rhesus monkey. *Vaccine*, *15*, 1872–1887.

Reise Sousa, C. (2006). Dendritic cells in a mature age. *Nature Reviews. Immunology*, *6*, 476–483.

Rezwan, M., Grau, T., Tschumi, A., & Sander, P. (2007). Lipoprotein synthesis in mycobacteria. *Microbiology*, *153*, 652–658.

Shang, E. S., Summers, T. A., & Haake, D. A. (1996). Molecular cloning and sequence analysis of the gene encoding LipL41, a surface-exposed lipoprotein of pathogenic *Leptospira* species. *Infection and Immunity*, *64*, 2322–2330.

Sung, W. H., Hsieh, S. Y., Lin, C. L., Leng, C. H., Liu, S. J., Chou, A. H., et al. (2010). Biochemical characterizations of E. coli expressed protective antigen Ag473 of *Neisseria meningitides* group B. *Vaccine*, *28*, 8175–8182.
Takeuchi, O., & Akira, S. (2010). Pattern recognition receptors and inflammation. *Cell*, *140*, 805–820.
Thoma-Uszynski, S., Stenger, S., Takeuchi, O., Ochoa, M. T., Engele, M., Sieling, P. A., et al. (2001). Induction of direct antimicrobial activity through mammalian toll-like receptors. *Science*, *291*, 1544–1547.
Tseng, C. L., & Leng, C. H. (2012). Influence of medium components on the expression of recombinant lipoproteins in *Escherichia coli*. *Applied Microbiology and Biotechnology*, *93*, 1539–1552.
von Heijne, G. (1989). The structure of signal peptides from bacterial lipoproteins. *Protein Engineering*, *2*, 531–534.
Yang, X., Izadi, H., Coleman, A. S., Wang, P., Ma, Y., Fikrig, E., et al. (2008). *Borrelia burgdorferi* lipoprotein BmpA activates pro-inflammatory responses in human synovial cells through a protein moiety. *Microbes and Infection*, *10*, 1300–1308.

CHAPTER FOUR

Perspective of Peptide Vaccine Composed of Epitope Peptide, CpG-DNA, and Liposome Complex Without Carriers

Younghee Lee*, Young Seek Lee†, Soo Young Cho‡, Hyung-Joo Kwon§,¶,1

*Department of Biochemistry, College of Natural Sciences, Chungbuk National University, Cheongju, South Korea
†Division of Molecular and Life Sciences, College of Science and Technology, Hanyang University, Ansan, South Korea
‡Laboratory of Developmental Biology and Genomics, College of Veterinary Medicine, Research Institute for Veterinary Science BK21, Program for Veterinary Science, Seoul National University, Seoul, South Korea
§Department of Microbiology, College of Medicine, Hallym University, Chuncheon, South Korea
¶Center for Medical Science Research, College of Medicine, Hallym University, Chuncheon, South Korea
[1]Corresponding author: e-mail address: hjookwon@hallym.ac.kr

Contents

Abstract

The magnitude and specificity of cell-mediated and humoral immunity are critically determined by peptide sequences; peptides corresponding to the B- or T-cell receptor

Advances in Protein Chemistry and Structural Biology, Volume 99
ISSN 1876-1623
http://dx.doi.org/10.1016/bs.apcsb.2015.03.004

epitopes are sufficient to induce an effective immune response if delivered properly. Therefore, studies on the screening and application of peptide-based epitopes have been done extensively for the development of therapeutic antibodies and prophylactic vaccines. However, the efficacy of immune response and antibody production by peptide-based immunization is too limited for human application at the present. To improve the efficacy of vaccines, researchers formulated adjuvants such as alum, water-in-oil emulsion, and Toll-like receptor agonists. They also employed liposomes as delivering vehicles to stimulate immune responses. Here, we review our recent studies providing a potent method of epitope screening and antibody production without conventional carriers. We adopted Lipoplex(O), comprising a natural phosphodiester bond CpG-DNA and a specific liposome complex, as an adjuvant. Lipoplex(O) induces potent stimulatory activity in humans as well as in mice, and immunization of mice with several peptides along with Lipoplex(O) without general carriers induces significant production of each peptide-specific IgG2a. Immunization of peptide vaccines against virus-associated antigens in mice has protective effects against the viral infection. A peptide vaccine against carcinoma-associated antigen and the peptide-specific monoclonal antibody has functional effects against cancer cells in mouse models. In conclusion, we improved the efficacy of peptide vaccines in mice. Our strategy can be applied in development of therapeutic antibodies or in defense against pandemic infectious diseases through rapid screening of potent B-cell epitopes.

1. INTRODUCTION

The immune system can recognize foreign antigens or altered self-antigens as a nonself and induce the cell-mediated immunity and humoral (antibody) immunity to defense our body. Innate immunity and adaptive immunity are another axis of immune system in terms of specialty. The major role players in the immune system include B cells, T cells, macrophages, dendritic cells (DCs), etc.

B cells have antigen-specific receptors on the cell surface and thereby can recognize native intact antigens. Presentation of antigens to B cells can induce cross-linking of the antigen-receptor molecules and result in the proliferation of antibody-producing B cells. The activated B cells secrete antibodies of IgM with low antigen affinity at first and then experience a series of events such as class switching, affinity maturation, and memory cell development. The necessity of second signal including cell–cell interaction and cytokines in this process was also documented (Mond, Lees, & Snapper, 1995; Sulzer & Perelson, 1997). For example, activated T helper (Th) cells provide signals necessary to amplify the immune response of B cells against the target antigen.

Different from B cells, T cells specifically recognize antigenic peptides which were processed and bound to the major histocompatibility complex (MHC) molecules in the antigen-presenting cells (APCs). The MHC molecules are cell surface glycoproteins to present antigenic peptides to T lymphocytes. MHC class I molecules display peptides derived from virus or altered self-antigen to cytotoxic T cells. Therefore, cytotoxic T lymphocytes (CTLs) that can detect tumor antigens bound to MHC class I complex have been shown to be potent effector cells to treat tumors. On the other hand, MHC class II molecules display peptides derived from exogenous antigens to helper T cells (Th cells). The MHC class II-binding peptides have been developed as preventive vaccines against various pathogens. Many investigators have studied peptide length bound to MHC class I and II molecules and the properties of peptide–MHC interaction to figure out common properties of the loaded peptides and to aid screening of T-cell epitopes (Jackson, Purcell, Fitzmaurice, Zeng, & Hart, 2002; Purcell et al., 2003; Tong, Tan, & Ranganathan, 2007).

Considering that B- or T-cell receptor epitopes are the peptides containing the minimal sequences necessary for immunomodulation of B and T cells, optimal selection of epitope peptides is an essential step for obtaining efficacious cell-mediated and humoral immunity *in vivo*. Surely, the peptides have to be delivered properly to elicit an effective immune response. Peptides can be used to induce a protective response against a particular pathogen such as malaria, influenza virus, hepatitis B, and HIV, to treat chronic diseases such as cancer, or to control autoimmune diseases (Ben-Yedidia & Arnon, 2007; Bijker, Melief, Offringa, & van der Burg, 2007; Schreiber, Raez, Rosenblatt, & Podack, 2010; Tam et al., 1990).

Epitope-based peptide vaccines to induce adaptive immunity have been extensively studied in various animal models. Peptides can be easily and promptly synthesized and the cost is inexpensive. Therefore, epitope-based peptide vaccines are considered as potentially useful prophylaxis for cancers and infectious diseases. However, peptide vaccines have poor immunogenicity in general, and the efficacy of peptide vaccines is limited to treat patients especially when used alone. To solve this problem and to efficiently activate immune response, peptide vaccines have to be improved by using potent immunostimulatory adjuvants.

Many types of vaccine adjuvants such as aluminum salt, oil emulsion, and TLR agonists have been tried since the 1970s, and now many investigators reported recent progress in preclinical trials and human trials (Black, Trent, Tirrell, & Olive, 2010; Mbow, De Gregorio, Valiante, & Rappuoli, 2010).

In this review, we will examine general adjuvants and then move to CpG-DNA and liposome. Then, we will present our recent research on peptide vaccines using a novel strategy to stimulate the innate immunity with a phosphodiester form of CpG-DNA and to facilitate delivery of vaccines using liposomes (Kwon et al., 2012; Rhee et al., 2012).

2. LICENSED ADJUVANTS

Vaccine adjuvants are defined as components to induce more potent immune response when added to antigen. To say in detail, adjuvants increase antibody titers, induce rapid immune response and long-lasting memory, reduce required antigen doses, and enhance immune response in the elderly and young children (Riese, Schulze, Ebensen, Prochnow, & Guzmán, 2013).

For long time, adjuvants have been developed and used empirically, and the most widely used adjuvants are insoluble aluminum salts (so-called alum). As alum has safety and efficacy records for about 100 years, alum is commercially used in a variety of vaccines against diphtheria, tetanus, pertussis, pneumococcus, hepatitis A and B, anthrax, *Haemophilus influenzae*, and human papilloma virus (Baylor, Egan, & Richman, 2002; Harrison, 1935; Hogenesch, 2013). The action mechanisms of alum seem complex and many factors simultaneously contribute to its effect. Alum was originally selected as a good adsorbent and this property endows injected sites with enhanced antigen uptake and stability (Goto & Akama, 1982; Goto et al., 1997). Alum also induces significant cellular infiltration into the injection sites and triggers a local proinflammatory reaction that further enables efficient processing and presentation of antigens (Kool et al., 2008; Morefield et al., 2005). However, alum cannot be the most effective adjuvant in every case: alum can also give negative effect on antigen adsorption (Aebig et al., 2007). Therefore, other potent adjuvants have been also investigated widely.

Emulsion adjuvants are known to be very effective for several decades; however, it could not be included in the vaccine formulation for long time because of nondegradable oil components. After development of biodegradable or biocompatible oils, oil-in-water emulsions have been licensed for prophylaxis in human (O'Hagan, Ott, Nest, Rappuoli, & Giudice, 2013). Typical examples are MF59 and AS03. MF9 is the squalene-based oil–water emulsion which was licensed in Europe for flu vaccines especially for the elderly and young children (Podda, 2001; Vesikari, Groth, Karvonen, Borkowski, & Pellegrini, 2009). AS03 is similar squalene-based oil-in-water emulsion containing the adjuvant alpha tocopherol. MF59 and AS03 have been licensed in Europe for pandemic flu vaccines and used for H1N1 pandemic flu vaccine in 2009 (Clark et al., 2009). However, an issue of side effect such as narcolepsy was raised for AS03 (Nohynek et al., 2012). The action mechanism of emulsion is considered to be the cellular recruitment and stimulation of the cells to produce cytokines and chemokines in the local injection sites rather than the depot effect to induce slow release of antigens (Calabro et al., 2011). Adjuvant effects of oil-in-water emulsions were also confirmed in the case of avian H5 pandemic flu vaccines (Gillard et al., 2013; Schwarz et al., 2009).

Another emerging group of vaccine adjuvant is the agonists of Toll-like receptors (TLRs). TLRs are expressed on various types of cells, including B cells, subset of T cells, monocytes, DCs, etc. (Iwasaki & Medzhitov, 2004). TLRs recognize common motifs called pathogen-associated molecular pattern (PAMP) in bacteria, fungi, viruses, and other pathogens (Akira, Uematsu, & Takeuchi, 2006). Since researchers reveal the critical importance of TLR signaling in the modulation of innate immunity as well as adaptive immunity against pathogens, TLR agonists including CpG-DNA, flagellin, and lipid became essential candidates of effective and safe vaccine adjuvants. TLR expression patterns are different depending on the specific cell types, and TLR agonists activate multiple TLR signaling pathways. Therefore, TLR agonists can effectively modulate immune responses of target cells (Black et al., 2010). For example, TLR agonists improve the efficacy of vaccine, reducing TCR-based selection thresholds and enhancing the magnitude and quality of memory T-cell response (Malherbe, Mark, Fazilleau, McHeyzer-Williams, & McHeyzer-Williams, 2008; Wille-Reece et al., 2006). Furthermore, the safety of TLR agonists as a component of vaccine was confirmed in clinical trials involving hepatitis B and malaria vaccines (Dupont et al., 2006; Ellis et al., 2009). Finally, the TLR4 agonist 3-O-desacyl-40-monophosphoryl lipid A (MPL) has been approved for human use in Europe (Casella & Mitchell, 2008). TLR agonists in combination with other adjuvant can further improve the efficacy; AS04, a combination adjuvant comprising MPL and alum, was licensed for HBV and HPV vaccines (Harper, 2009; Thoelen et al., 1998).

3. CpG-DNA: ADJUVANT

CpG-DNA represents synthetic oligodeoxynucleotides (ODNs) and bacterial DNA including unmethylated CpG motifs (Krieg et al., 1995).

The frequency of natural CpG-DNA occurrence in bacterial DNA is much higher than that of the eukaryotic chromosome, and CpG-DNA has immunostimulatory activity as a typical example of PAMP molecules. Therefore, recognition of CpG-DNA and activation of immune system contribute to effective removal of pathogens. Among several TLRs, TLR9 recognizes CpG-DNA (Hemmi et al., 2000). TLR9 is expressed in the multiple cell types including B cells, monocytes, natural killer cells, dendritic cells, and macrophages (Bauer et al., 2001; Hemmi et al., 2000). TLR9 is located on the membrane of endosomal compartments in the cells; therefore, CpG-DNA is first internalized and then interacts with TLR9 to stimulate the target cells (Brinkmann et al., 2007; Kim, Brinkmann, Paquet, & Ploegh, 2008). TLR9 signaling induced by CpG-DNA stimulates the expression of proinflammatory cytokines (e.g., TNF-α, IL-1, IL-6, IL-12, IFN-α), chemokines (e.g., MIP-2, MCP-1, RANTES, IP-10), MHC class II molecules, and costimulatory molecules (CD40, CD80, CD83, and CD86) (Kim, Lee, & Kwon, 2013; Krieg, 2002; Kwon & Kim, 2003; Kwon, Lee, Yu, Han, & Kim, 2003; Lee, Lee, Kim, & Kwon, 2006; Lee, Lee, Kwon, & Kim, 2005; Park, Kim, Lee, & Kwon, 2013). Therefore, CpG-DNA has gained attention as an effective vaccine adjuvant (Carson & Raz, 1997; Chu, Targoni, Krieg, Lehmann, & Harding, 1997). Animal studies and Phase I–III clinical trials demonstrated that CpG-DNA satisfies safety issue and provides good efficacy as a vaccine adjuvant (Bode, Zhao, Steinhagen, Kinjo, & Klinman, 2011; Brody et al., 2010; Karbach et al., 2010).

Most synthetic ODNs used for studies can be classified to four groups such as K-type (or B-type), D-type (or A-type), C-type, and P-type depending on the content of phosphodiester/phosphorothioate in the backbone, frequency of CpG sequences, and properties of surrounding sequences (Hartmann et al., 2003; Marshall et al., 2003; Verthelyi, Ishii, Gursel, Takeshita, & Klinman, 2001). The immunostimulatory effects of CpG-DNAs on immune cells are various depending on the specific CpG-DNA types (Hartmann & Krieg, 2000; Mutwiri, Nichani, Babiuk, & Babiuk, 2004; Samulowitz et al., 2010). All of these ODNs include phosphorothioate backbone. Therefore, we call them phosphorothioate backbone-modified CpG-DNA (PS-ODN). The sulfurs replacing the nonbridging oxygen atoms in the backbone stabilize the ODNs. PS-ODN is able to dramatically enhance the CpG-DNA-induced immune responses by resistance to nuclease activity and efficacious delivery into the cells (Stein, Subasinghe, Shinozuka, & Cohen, 1988; Zhao et al.,

1993; Zhao, Waldschmidt, Fisher, Herrera, & Krieg, 1994). However, PS-ODN induces severe side effects such as arthritis, lymphoid follicle destruction, and transient splenomegaly in a CG sequence-dependent and backbone modification-dependent manner (Deng, Nilsson, Verdrengh, Collins, & Tarkowski, 1999; Heikenwalder et al., 2004; Sparwasser et al., 1999). Furthermore, we previously reported production of PS-ODN-specific IgM in PS-ODN-treated mice (Kim et al., 2009).

4. PHOSPHODIESTER BOND CpG-DNA

Different from the PS-ODN described above, there is another group of CpG-DNA which is composed of the natural phosphodiester bond oligonucleotides (PO-ODNs). PO-ODN is a regulator of the immune response and an adjuvant inducing Ag-driven Th1 responses without severe side effects. Therefore, we and other investigators have developed PO-ODN, as a natural counterpart of PS-ODN, which can induce optimal innate immune responses (Brown, Estes, Chantler, Kegerreis, & Suarez, 1998; Lee, Jung, et al., 2006).

To isolate the sequences of bacterial DNA that have immunostimulatory effects, we analyzed the genomic DNA sequences of *Mycobacterium bovis* and *Escherichia coli* with computer and selected candidates of 20 base length PO-ODNs that contain CpG motifs. Our experimental analysis proved that the PO-ODN, specifically MB-ODN 4531(O), activates the transcription factor NF-κB. Most importantly, use of MB-ODN 4531(O) or phosphorothioate backbone-modified MB-ODN 4531(S) as an adjuvant induced a much higher level of IgG2a than the case of incomplete Freund's adjuvant (IFA) which potently induces IgG1 production, suggesting that MB-ODN 4531 contributes to Th1 differentiation in the immune system. We also confirmed that MB-ODN 4531(O) regulates expression of Th1 cytokines such as IFN-γ and IL-12 in the innate immune system without severe side effects (Choi, Lee, Kwon, & Kim, 2006; Kim, Jung, Lee, & Kwon, 2011; Lee, Jung, et al., 2006). However, PO-ODN has very weak immunomodulatory activity compared to PS-ODN. Therefore, to develop PO-ODN as an effective vaccine adjuvant, there should be a strategy to increase the immunomodulatory activity of PO-ODN. Another problem is that the activity is found in mouse cells but not in human cells (Liang, Nishioka, Reich, Pisetsky, & Lipsky, 1996). However, PO-ODN encapsulated in cationic liposomes such as lipofectin and *N*-[1-(2,3-dioleoyloxy)propyl]-*N*,*N*,*N*-trimethylammonium methylsulfate (DOTAP)

(Magnusson, Tobes, Sancho, & Pareja, 2007; Yasuda et al., 2005) induced an effective immune response in human cells. Therefore, we postulated that PO-ODN in combination with liposomes can be an effective strategy to improve the potency of PO-ODN.

5. LIPOSOMES: ADJUVANTS

Liposomes protect antigens from the environment and are potent vehicles for delivering antigens to APCs. Liposomes in the vaccine also regulate the local tissue distribution, retention, and cell trafficking in the injected place (Oussoren et al., 1998). Therefore, liposomes are known to enhance antibody production and CTL responses to coadministrated antigens (Bhowmick, Mazumdar, Sinha, & Ali, 2010; Chang, Choi, Cheong, & Kim, 2001; Chikh & Schutze-Redelmeier, 2002; Gursel, Gursel, Ishii, & Klinman, 2001).

The features of immune responses enhanced by liposomes are different according to the lipid composition, charge, size, and the location of antigens or adjuvants (Carstens et al., 2011). The pH-sensitive liposomes such as phosphatidyl-β-oleoyl-γ-palmitoyl ethanolamine (DOPE):cholesterol hemisuccinate (CHEMS) enhance the efficiency of antigen delivery to the cytosol and activate CTL responses (Chang et al., 2001). Sterical stabilization of cationic liposomes with the aid of polyethylene glycol increases the uptake of antigen in immune cells as can be seen for DOPE:polyethylene glycol (Gursel et al., 2001). Cationic liposomes such as lipofectamine, 3β-[*N*-(*N*′,*N*′-dimethylaminoethane)-carbamoyl]cholesterol hydrochloride (DC-Chol), DC-Chol:DOPE, and phosphatidylcholine:stearylamine:cholesterol improve CTL response and antibody production (Chikh & Schutze-Redelmeier, 2002). However, it is likely that the charge of liposome is not a single critical factor; for example, ovalbumin injected with negative 1,2-dioleoyl-*sn*-glycero-3-phosphatidic acid liposomes was as immunogenic as ovalbumin administrated with positive 1,2-dioleoyl-3-trimethyl ammonium propane liposomes (Yanasarn, Sloat, & Cui, 2011). It was also reported that small liposome induces Th1 response, while large liposome induces Th2 response (Badiee et al., 2012). Depending on the solubility of antigens in water or lipid, the location of antigen can be assigned to entrapment or deposition in the lipid. Furthermore, the administrated methods such as adsorption, encapsulation, and chemical cross-linking affect the location of antigens and the magnitude of immune responses. It was reported that adsorbed hemagglutinin (HA) was more immunogenic than encapsulated one. Furthermore, other components such as cholesterol or adjuvants further enhance the responses (Barnier-Quer, Elsharkawy, Romeijn, Kros, & Jiskoot, 2012, 2013).

6. ADJUVANTS FOR PEPTIDE VACCINE

B-cell epitopes bind to B-cell receptors, and T-cell epitopes associates with MHC to make a complex that can be recognized by T-cell receptors. Therefore, peptide vaccines based on the epitopes have been studied for the past 30 years, and the capability of the peptide vaccines to induce and regulate immune responses has been proved. The production process of peptide vaccine is relatively easier than conventional antigen preparation. Peptide can be easily synthesized, and multiple peptide epitopes in a single peptide can be designed and tested. The solubility and stability of the peptides can be enhanced by incorporation of carbohydrates (Jackson et al., 2002). Therefore, peptide vaccines are potentially useful as prophylaxis for a variety of diseases including cancers and infectious diseases (Naz & Dabir, 2007). However, there is a critical challenge to overcome: the low immunogenicity of peptide vaccines compared to whole pathogen vaccines.

To improve the efficacy of peptide vaccines, there are several types of adjuvants available. The most commonly used adjuvants are water-in-oil emulsions such as IFA or the human equivalent Montanide ISA-51. Historically, the first peptide vaccine inducing a T-cell response in mice was composed of 15-mer peptide derived from NP protein of LCMV in combination with IFA (Aichele, Hengartner, Zinkernagel, & Schulz, 1990). The water-in-oil emulsions are strong adjuvants and they majorly contribute to prolonged release of antigen as a depot and also improve antigen presentation by APCs. However, there is limitation of these adjuvants for application in clinics. IFA induced side effects such as swelling, redness, and increased skin temperature owing to local lesions on the site of injection (Kenter et al., 2009). It is speculated that the depot effect can induce tolerance if the ratio of peptide and adjuvant was not properly maintained *in vivo* (Welters et al., 2007). Montanide ISA-51 was used as an adjuvant for clinical trials involving HIV peptides and malaria surface proteins; however, the trials were terminated because of severe local and systemic adverse effects (Graham et al., 2010; Wu et al., 2008).

As the route of administration and the specific formulation of the peptide vaccine can affect the features and magnitude of the immune response, the peptide has to be presented using multifunctional and efficacious delivery system (Black et al., 2010). Therefore, alternative group of adjuvants could be the use of vesicular delivery systems including liposomes, virosomes, virus-like-particles, and nanoparticles (Foged, Hansen, & Agger, 2012). TLR agonists such as CpG-DNA, flagellin, and MPL can be incorporated with the peptide delivery system. For example, BLP25 liposome vaccine (L-BLP25) was prepared by lyophilization with lipopeptide, MPL, and three lipids (cholesterol, dimyristoyl phosphatidylglycerol, and dipalmitoyl phosphatidylcholine). L-BLP25 was proved to be safe and effective against nonsmall-cell lung cancer (Sangha & Butts, 2007). Antigenic peptides can be also displayed on the surface of virus-like particles. Preclinical and clinical studies using this strategy have been done, and virus-like particles-conjugated influenza vaccines, HBV and HPV vaccines, were approved for human use (Grgacic & Anderson, 2006; Herzog et al., 2009). Nanoparticles with proper size can be conjugated with antigenic peptides and directly can be used as an effective vaccine suggesting that the size is a critical factor to stimulate DCs. When multiple peptides from the foot-and-mouth virus were conjugated to nanobeads of 45 nm polystyrene and immunized, humoral and cell-mediated immune responses were induced without additional adjuvants (Greenwood et al., 2008).

7. PEPTIDE VACCINE COMPOSED OF EPITOPE PEPTIDE, CpG-DNA, AND LIPOSOME COMPLEX WITHOUT CARRIERS

As we decided to use phosphodiester bond CpG-DNA (PO-ODN) as an adjuvant without side effects and peptide as a specific antigen, we have to find an efficacious strategy to maximize the potency of the vaccine. Therefore, we screened best formulation of liposome complex and then applied the formulation to peptide vaccines.

To enhance the potency of PO-ODN in human cells, we prepared and tested several different compositions of liposome complexes and found that DOPE/CHEMS (1:1 ratio) was the best choice. The natural CpG-DNA encapsulated with the DOPE/CHEMS complex (Lipoplex(O)) increased expression of cytokines such as IL-6, IL-12, and IFN-γ in human cells as well as in mouse cells. This process accompanied improved intracellular uptake of CpG-DNA and TLR9/MyD88-mediated cellular activation. Furthermore, injection of Lipoplex(O) increased the levels of cytokine expression in mouse serum, which was TLR9 dependent and CpG sequence dependent (Kim, Kwon, Ahn, et al., 2011; Kim et al., 2010).

We applied the liposome formula to epitope peptide vaccine. When the CpG-DNA–liposome coencapsulated with epitope peptide was injected into the mice, the vaccine caused the induction of epitope peptide-specific antibody production. First, we applied a complex of epitope peptide and Lipoplex(O) system to the screening of B-cell epitopes against proteins of infectious pathogens and evaluated the efficacy of the prophylactic vaccine with a complex of epitope peptide and Lipoplex(O). We successfully screened B-cell epitopes against virus infection-associated proteins: Hepatitis C virus envelope protein (HCV-E) (Kim, Kwon, & Lee, 2011), attachment glycoprotein G and fusion protein F of human respiratory syncytial virus (HRSV-G and HRSV-F) (Kim, Kwon, & Lee, 2011; Park et al., 2014), and HA protein of influenza A virus (Rhee et al., 2012) by immunization with complexes of several epitope peptides and Lipoplex(O) without carriers. Second, we screened the B-cell epitope of the hepatocellular carcinoma (HCC)- and colon cancer-specific transmembrane 4 superfamily member 5 (TM4SF5) protein and evaluated the efficacy of the prophylactic and therapeutic cancer vaccines with a complex of epitope peptide and Lipoplex(O) (Kwon, Kim, Kim, et al., 2013; Kwon et al., 2012; Kwon, Kim, et al., 2014; Kwon, Kim, Park, et al., 2013).

7.1 Antibody Production with Epitope Peptide, CpG-DNA, and Liposome Complex Without Carriers

When BALB/c mice immunized i.p. with complexes of predicted epitope peptide and PO-ODN encapsulated in DOPE:CHEMS liposomes (Lipoplex(O)), the mice produced significantly higher amount of epitope peptide-specific IgG than the control mice. However, the production of peptide-specific IgG was much lower by immunization with a complex of epitope peptide and PO-ODN coencapsulated in other liposomes, which suggests the superior adjuvant effect of Lipoplex(O). The most effective antibody production was observed with DOPE:CHEMS (1:1 ratio) (Kim, Kwon, Rhee et al., 2011).

The adjuvant effect of Lipoplex(O) in the peptide-specific IgG production was dependent on the CG sequence of PO-ODN, but not on backbone modification (PO-ODN vs. PS-ODN). The complex of epitope peptide and Lipoplex(O) induced production of peptide-specific IgG in a TLR9-dependent manner. Importantly, the complex of epitope peptide and Lipoplex(O) was significantly useful in producing epitope-specific IgG2a isotype than IgG1. Production of IgG and IgG2a in response to a complex of epitope peptide and Lipoplex(O) immunization was reduced in

STAT4-knockout mice. In contrast, STAT6 knockout had no influence on IgG production. Considering that STAT4 is involved in the differentiation of T cells into Th1 cells, it is likely that Th1-dominated humoral response is induced by complex of epitope peptide and Lipoplex(O). Next, IgG production by the peptide vaccine required $CD4^+$ T cells. $CD4^+$-depleted mice produced much less peptide-specific IgG. Furthermore, the production of peptide-specific IgG and IgG2a was also significantly decreased in MHC class II-knockout mice (Kwon et al., 2012). Taken together, antibody production induced by the complex of epitope peptide and Lipoplex(O) is Th1 dominated, $CD4^+$ T cell dependent, MHC dependent, and TLR9 dependent.

7.2 Prevention of Influenza A Virus Infection

Influenza A viruses are pathogenic in birds and humans, which have gained attention because outbreak of influenza pandemic is an important cause of human diseases associated with respiratory tract disease and mortality (Peiris, de Jong, & Guan, 2007). Recently, pandemic influenza A subtypes including H1N1, H5N1, H9N2, and H7N7 are verified as origin of birds and/or humans diseases (Abdel-Ghafar et al., 2008; Centers for Disease Control and Prevention (CDC), 1997; Fouchier et al., 2004; Garten et al., 2009; Peiris et al., 1999). Investigators have performed diverse researches to develop prophylactic vaccines and therapeutics against various influenza viruses. Inactivated whole-influenza virus vaccines have been conventionally developed and commercially available in the marketplaces (Neuzil et al., 2006). Live attenuated vaccines have also been developed and the live attenuated virus is also commercially available in some marketplaces (Fodor et al., 1999). Although these conventional virus-based influenza vaccines are commonly tried in the clinic, the vaccine productions need amounts of specific-pathogen-free embryonated chicken eggs. Further, the limited production of conventional virus-based influenza vaccines could not promptly cope with the influenza pandemic outbreak situation. To overcome the limitation of virus-based influenza vaccines, epitope-based peptide vaccines have been investigated.

Influenza HA is a surface glycoprotein (HA0), which is cleaved to two polypeptides, HA1 and HA2, and essential to the entry process (Wilson, Skehel, & Wiley, 1981). Therefore, HA protein has been studied as a target for influenza immunotherapy. To select the B-cell epitopes that originated from HA protein of avian influenza A type of H5N1 Viet Nam strain 2004,

BALB/c mice were immunized with the complex of each candidate epitope from HA1 and Lipoplex(O). The BALB/c mice induced production of IgG specific to hH5N1 HA58 and hH5N1 HA233 peptide more markedly than other peptides (as shown in Table 1). The complex of hH5N1 HA epitope peptide and Lipoplex(O) significantly enhanced production of peptide-specific IgG in a CG sequence of PO-ODN-dependent manner. Further, the immunization with a complex of epitope and Lipoplex(O) produced a much higher level of IgG2a than IgG1 in the mice, suggesting that Lipoplex(O) induces a Th1-dominated humoral response. To evaluate the prophylactic effect of a vaccine containing a complex of hH5N1 HA233 and Lipoplex(O), BALB/c mice were challenged with the recombinant H5N1 virus 10 days after the vaccination. While all of the mice immunized with a vaccine containing a complex of hH5N1 HA233 and Lipoplex(O) survived, all the mice of the negative control groups died. The mice immunized with the hH5N1 HA233 peptide vaccine had regained body weight and recovered from pulmonary lesions consisting of severe necrotizing bronchitis and severe histiocytic alveolitis. To examine

Table 1 Selected B-Cell Epitope by Immunization with Complexes of Candidate Epitope Peptide and Lipoplex(O)

Sources	Epitope Sequences	Abbreviations	References
Influenza A virus HA protein	^{233}IATRSKVNGQSGRM^{246}	hH5N1 HA233	Rhee et al. (2012)
RSV F protein	^{346}AGSVSFFPQAETCKV^{360} ^{393}CKIMTSKTDVSSSVI^{407}	F7 F9	Park et al. (2014)
RSV G protein	^{1}MSKHKNQRTARTLEKTWD^{18}	HRSV-G1	Kim, Kwon, and Lee (2011)
HCV-E protein	^{202}CFRKHPEATYSR^{213}	HCV-E2 202	Kim, Kwon, and Lee (2011)
TM4SF5	^{138}NRTLWDRCEAPPRV^{151}	TM4SF5R2-3	Kim, Kwon, Rhee, et al. (2011)

Candidate epitopes were selected from indicated proteins based on hydrophilicity, hydrophobicity, secondary structure, antigenicity index, and amphipathicity (http://tools.immuneepitpoe.org/main/index.html). BALB/c mice were injected intraperitoneally with candidate epitope peptide and PO-ODN encapsulated in DOPE:CHEMS liposomes (Lipoplex(O)).
HA, hemagglutinin; HCV, hepatitis C virus; PO-ODN, natural phosphodiester bond CpG-DNA, namely MB-ODN 4531(O), derived from *M. bovis* genomic DNA; RSV, respiratory syncytial virus; TM4SF5, tetraspanin transmembrane 4 superfamily member 5.

whether immunization with the hH5N1 HA233 peptide vaccine induces a prophylactic effect on other influenza subtype, the BALB/c mice were challenged with the mouse-adapted H1N1 virus (maA/WSN/1933) after vaccination and 50% of the mice survived (Rhee et al., 2012). Thus, further studies are necessary to evaluate whether the complex of hH5N1 HA233 and Lipoplex(O) has prophylactic effect on the challenge of various influenza A subtypes. Such studies will allow fine-tuning and improvement of a potential vaccine to defense against pandemic influenza.

7.3 Prevention of Respiratory Syncytial Virus Infection

Respiratory syncytial virus (RSV) is a common disease virus infecting lung epithelium and leading to seriously life-threatening acute bronchiolitis and pneumonia in infants and young children (Harris & Werling, 2003; Kong et al., 2003). Until now, RSV vaccine is not available in the marketplace. Although formalin-inactivated whole RSV vaccine (FI-RSV) has been conventionally developed, applications of the FI-RSV were not approved because FI-RSV induced Th2-type hyperimmune responses (Connors et al., 1994; Haynes, Jones, Barskey, Anderson, & Tripp, 2003). To overcome this problem, we used Lipoplex(O) as an adjuvant to induce a Th1-dominated humoral response in animal experiments. Although a complex of UV-irradiated RSV (UV-RSV) and Lipoplex(O) induced a Th1-dominated humoral response producing majorly IgG2a, the UV-RSV vaccine did not induce vaccination effects against RSV infection (Park et al., 2014). It is likely that the whole virus vaccine is not a proper antigen for our system. Therefore, we evaluated the efficacy of peptide vaccine using our strategy.

RSV fusion (F) protein is found on the envelope of the RSV and mediates fusion and formation of syncytia. Therefore, F protein has gained attention as a target for RSV vaccination. To develop the epitope-based peptide vaccine, the B-cell epitopes originated from F protein of RSV A strain were selected by immunization with the complex of each candidate epitope from F protein and Lipoplex(O). Among the candidates, F7 and F9 peptide-containing peptide vaccines induced most significant production of each peptide-specific IgG and IgG2a (as shown in Table 1). To evaluate the prophylactic efficacy of a vaccine containing a complex of F7 (or F9) epitope peptide and Lipoplex(O), BALB/c mice were challenged with the recombinant RSV A2 strain 10 days after the vaccination. Based on lung histopathology and mucus clearance in the lung, the F7 and F9 peptide vaccine induced prophylactic effect on RSV (Park et al., 2014). Thus, further studies on the efficacy of the peptide vaccines against multiple RSV strains are needed to optimize and improve the peptide vaccine formula.

7.4 Cancer Vaccine Composed of Epitope Peptide, CpG-DNA, and Liposome Complex Without Carriers

TM4SF5 has been implicated in HCC, colon cancer, and pancreatic cancer (Muller-Pillasch et al., 1998). The expression of TM4SF5 induced morphological elongation, epithelial–mesenchymal transition, uncontrolled growth of human HCC cells *in vitro*, and tumor formation *in vivo* (Lee et al., 2008). Therefore, TM4SF5 protein has gained attention as a target for HCC, colon cancer, and pancreatic cancer therapy.

The B-cell epitopes of hTM4SF5 protein were selected by immunization with complexes of each predicted hTM4SF5 peptide and Lipoplex(O) (as shown in Table 1). The production of peptide-specific IgG was most enhanced by a complex comprising hTM4SF5R2-3 peptide and Lipoplex(O). Mouse HCC cells implanted into BALB/c mice grew continuously into a tumor mass. Vaccination with a complex of hTM4SF5R2-3 peptide and Lipoplex(O) induced production of hTM4SF5R2-3 peptide-specific antibody and showed delayed tumor formation. However, there was no significant vaccination effect of the peptide vaccine in the TLR9-knockout mice suggesting that CpG-DNA is critical for eliciting a protective immune response with the peptide vaccine. Interestingly, the tumor growth was lower in the TLR9 −/− mice than in wild-type mice (Kwon et al., 2012; Kwon, Kim, Park, et al., 2013).

TM4SF5 protein is also expressed in the mouse colon cancer cell line CT-26. The vaccination with the TM4SF5 peptide vaccine induces production of TM4SF5-specific antibodies and an effective protective immune response against colon tumor established by implantation of CT-26 cells. Therefore, TM4SF5 can be an efficacious target for anticancer therapy against HCC as well as colon cancers (Kwon, Kim, Kim, et al., 2013; Kwon, Kim, et al., 2014). Considering that pancreatic cancer tissues also express TM4SF5, evaluation of the peptide vaccine in pancreatic cancer model is warranted.

8. PERSPECTIVES

The immunostimulatory ability of natural phosphodiester CpG-DNA (PO-ODN) to induce Th1-associated immune responses suggests that they may be promising adjuvants for vaccines to prevent cancers and infectious diseases. Especially, PO-ODN may be more optimal than phosphorothioate-modified CpG-DNA (PS-ODN) because PO-ODN modulates innate immune responses without severe side effects. The effect of PO-ODN on innate immune response can be upregulated by encapsulation in a DOPE:CHEMS complex. When the epitope peptide was coencapsulated with Lipoplex(O) without carriers, epitope peptide-specific IgG production was significantly induced. A complex of the epitope peptide and Lipoplex(O) capable of inducing IgG production had prophylactic effects against cancer and infectious diseases; therefore, this platform could be applied for general peptide vaccines. Although we know that our peptide vaccines activate Th1 response to induce peptide-specific IgG production, further investigations are required to verify the mechanisms involved. For example, we screened and used only B-cell epitopes; however, the ability of peptide vaccine is $CD4^+$ T cell dependent. Furthermore, detailed studies on the immunological responses *in vivo* including local immune cell infiltration, retention, cytokine profiles, T-cell activity, and memory response are also necessary. To develop prophylaxis and hopefully therapeutics for human use, animal and clinical studies on the safety and efficacy of the peptide vaccines have to be performed. For application to viral disease in humans and animals, additional animal studies to optimize the peptide vaccine formula have to be done with diverse viral strains. Even though we have a long way to go to develop safe and efficacious peptide vaccine, we believe that our novel strategy contributes to a more effective screening of potent B-cell epitopes at the present time and can be applied for the development of therapeutic antibodies in a near future. For example, we produced a monoclonal antibody by immunization with a complex of a hTM4SF5 epitope peptide and Lipoplex(O) and recently investigated that the monoclonal antibody can inhibit the growth of HCC and colon cancer established by implantation of cancer cells in mice (Kim et al., 2014; Kwon, Choi, et al., 2014). Our overall results show that complexes of epitope and Lipoplex(O) without carriers are useful for B-cell epitope screening, peptide vaccines, and peptide-specific antibody production for antibody-based cancer immunotherapy.

ACKNOWLEDGMENTS

This work was supported by grants from National Research Foundation of Korea (NRF) funded by the Ministry of Science, ICT & Future Planning in the Republic of Korea (2013R1A2A2A03067981, 2013M3A9A9050126, and 2009-0093812) to H.-J.K. and a grant from NRF (2012R1A5A1B68671075) to Y.S.L.

REFERENCES

Abdel-Ghafar, A. N., Chotpitayasunondh, T., Gao, Z., Hayden, F. G., Nguyen, D. H., de Jong, M. D., et al. (2008). Update on avian influenza A (H5N1) virus infection in humans. *The New England Journal of Medicine, 358*, 261–273.

Aebig, J. A., Mullen, G. E., Dobrescu, G., Rausch, K., Lambert, L., Ajose-Popoola, O., et al. (2007). Formulation of vaccines containing CpG oligonucleotides and alum. *Journal of Immunological Methods, 323*, 139–146.

Aichele, P., Hengartner, H., Zinkernagel, R. M., & Schulz, M. (1990). Antiviral cytotoxic T cell response induced by in vivo priming with a free synthetic peptide. *The Journal of Experimental Medicine, 171*, 1815–1820.

Akira, S., Uematsu, S., & Takeuchi, O. (2006). Pathogen recognition and innate immunity. *Cell, 124*, 783–801.

Badiee, A., Khamesipour, A., Samiei, A., Soroush, D., Shargh, V., Kheiri, M., et al. (2012). The role of liposome size on the type of immune response induced in BALB/C mice against leishmaniasis: rgp63 as a model antigen. *Experimental Parasitology, 132*, 403–409.

Barnier-Quer, C., Elsharkawy, A., Romeijn, S., Kros, A., & Jiskoot, W. (2012). Cationic liposomes as adjuvants for influenza hemagglutinin: More than charge alone. *European Journal of Pharmaceutics and Biopharmaceutics, 81*, 294–302.

Barnier-Quer, C., Elsharkawy, A., Romeijn, S., Kros, A., & Jiskoot, W. (2013). Adjuvant effect of cationic liposomes for subunit influenza vaccine: Influence of antigen loading method, cholesterol and immune modulators. *Pharmaceutics, 5*, 392–410.

Bauer, S., Kirschning, C. J., Häcker, H., Redecke, V., Hausmann, S., Akira, S., et al. (2001). Human TLR9 confers responsiveness to bacterial DNA via species-specific CpG motif recognition. *Proceedings of the National Academy of Sciences of the United States of America, 98*, 9237–9242.

Baylor, N. W., Egan, W., & Richman, P. (2002). Aluminum salts in vaccines—US perspective. *Vaccine, 20*, S18–S23.

Ben-Yedidia, T., & Arnon, R. (2007). Epitope-based vaccine against influenza. *Expert Review of Vaccines, 6*, 939–948.

Bhowmick, S., Mazumdar, T., Sinha, R., & Ali, N. (2010). Comparison of liposome based antigen delivery systems for protection against *Leishmania donovani*. *Journal of Controlled Release, 141*, 199–207.

Bijker, M. S., Melief, C. J., Offringa, R., & van der Burg, S. H. (2007). Design and development of synthetic peptide vaccines: Past, present and future. *Expert Review of Vaccines, 6*, 591–603.

Black, M., Trent, A., Tirrell, M., & Olive, C. (2010). Advances in the design and delivery of peptide subunit vaccines with a focus on toll-like receptor agonists. *Expert Review of Vaccines, 9*, 157–173.

Bode, C., Zhao, G., Steinhagen, F., Kinjo, T., & Klinman, D. M. (2011). CpG DNA as a vaccine adjuvant. *Expert Review of Vaccines, 10*, 499–511.

Brinkmann, M. M., Spooner, E., Hoebe, K., Beutler, B., Ploegh, H. L., & Kim, Y. M. (2007). The interaction between the ER membrane protein UNC93B and TLR3, 7, and 9 is crucial for TLR signaling. *The Journal of Cell Biology, 177*, 265–275.

Brody, J. D., Ai, W. Z., Czerwinski, D. K., Torchia, J. A., Levy, M., Advani, R. H., et al. (2010). *In situ* vaccination with a TLR9 agonist induces systemic lymphoma regression: A phase I/II study. *Journal of Clinical Oncology, 28*, 4324–4332.

Brown, W. C., Estes, D. M., Chantler, S. E., Kegerreis, K. A., & Suarez, C. E. (1998). DNA and a CpG oligonucleotide derived from *Babesia bovis* are mitogenic for bovine B cells. *Infection and Immunity, 66*, 5423–5432.

Calabro, S., Tortoli, M., Baudner, B. C., Pacitto, A., Cortese, M., O'Hagan, D. T., et al. (2011). Vaccine adjuvants alum and MF59 induce rapid recruitment of neutrophils and monocytes that participate in antigen transport to draining lymph nodes. *Vaccine, 29*, 1812–1823.

Carson, D. A., & Raz, E. (1997). Oligonucleotide adjuvants for T helper 1 (Th1)-specific vaccination. *The Journal of Experimental Medicine, 186*, 1621–1622.

Carstens, M. G., Camps, M. G., Henriksen-Lacey, M., Franken, K., Ottenhoff, T. H., Perrie, Y., et al. (2011). Effect of vesicle size on tissue localization and immunogenicity of liposomal DNA vaccines. *Vaccine, 29*, 4761–4770.

Casella, C. R., & Mitchell, T. C. (2008). Putting endotoxin to work for us: Monophosphoryl lipid A as a safe and effective vaccine adjuvant. *Cellular and Molecular Life Sciences, 65*, 3231–3240.

Centers for Disease Control and Prevention (CDC). (1997). Isolation of avian influenza A (H5N1) viruses from humans-Hong Kong, May-December 1997. *MMWR. Morbidity and Mortality Weekly Report, 46*, 1204–1207.

Chang, J. S., Choi, M. J., Cheong, H. S., & Kim, K. (2001). Development of Th1-mediated CD8+ effector T cells by vaccination with epitope peptides encapsulated in pH-sensitive liposomes. *Vaccine, 19*, 3608–3614.

Chikh, G., & Schutze-Redelmeier, M. P. (2002). Liposomal delivery of CTL epitopes to dendritic cells. *Bioscience Reports, 22*, 339–353.

Choi, Y. J., Lee, K. W., Kwon, H. J., & Kim, D. S. (2006). Identification of immunostimulatory oligodeoxynucleotides from *Escherichia coli* genomic DNA. *Journal of Biochemistry and Molecular Biology, 39*, 788–793.

Chu, R. S., Targoni, O. S., Krieg, A. M., Lehmann, P. V., & Harding, C. V. (1997). CpG oligodeoxynucleotides act as adjuvants that switch on T helper 1 (Th1) immunity. *The Journal of Experimental Medicine, 186*, 1623–1631.

Clark, T. W., Pareek, M., Hoschler, K., Dillon, H., Nicholson, K. G., Groth, N., et al. (2009). Trial of 2009 influenza A (H1N1) monovalent MF59-adjuvanted vaccine. *The New England Journal of Medicine, 361*, 2424–2435.

Connors, M., Giese, N. A., Kulkarni, A. B., Firestone, C. Y., Morse, H. C., 3rd, & Murphy, B. R. (1994). Enhanced pulmonary histopathology induced by respiratory syncytial virus (RSV) challenge of formalin-inactivated RSV-immunized BALB/c mice is abrogated by depletion of interleukin-4 (IL-4) and IL-10. *Journal of Virology, 68*, 5321–5325.

Deng, G. M., Nilsson, I. M., Verdrengh, M., Collins, L. V., & Tarkowski, A. (1999). Intra-articularly localized bacterial DNA containing CpG motifs induces arthritis. *Nature Medicine, 5*, 702–705.

Dupont, J., Altclas, J., Lepetic, A., Lombardo, M., Vázquez, V., Salgueira, C., et al. (2006). A controlled clinical trial comparing the safety and immunogenicity of a new adjuvanted hepatitis B vaccine with a standard hepatitis B vaccine. *Vaccine, 24*, 7167–7174.

Ellis, R. D., Mullen, G. E., Pierce, M., Martin, L. B., Miura, K., Fay, M. P., et al. (2009). A Phase 1 study of the blood-stage malaria vaccine candidate AMA1-C1/Alhydrogel with CPG 7909, using two different formulations and dosing intervals. *Vaccine, 27*, 4104–4109.

Fodor, E., Devenish, L., Engelhardt, O. G., Palese, P., Brownlee, G. G., & García-Sastre, A. (1999). Rescue of influenza A virus from recombinant DNA. *Journal of Virology, 73*, 9679–9682.

Foged, C., Hansen, J., & Agger, E. M. (2012). License to kill: Formulation requirements for optimal priming of CD8(+) CTL responses with particulate vaccine delivery systems. *European Journal of Pharmaceutical Sciences, 45*, 482–491.

Fouchier, R. A., Schneeberger, P. M., Rozendaal, F. W., Broekman, J. M., Kemink, S. A., Munster, V., et al. (2004). Avian influenza virus A (H7N7) associated with human conjunctivitis and a fatal case of acute respiratory distress syndrome. *Proceedings of the National Academy of Sciences of the United States of America, 101*, 1356–1361.

Garten, R. J., Davis, C. T., Russell, C. A., Shu, B., Lindstrom, S., Balish, A., et al. (2009). Antigenic and genetic characteristics of swine-origin 2009 A(H1N1) influenza viruses circulating in humans. *Science, 325*, 197–201.

Gillard, P., Caplanusi, A., Knuf, M., Roman, F., Walravens, K., Moris, P., et al. (2013). An assessment of prime-boost vaccination schedules with AS03A-adjuvanted prepandemic H5N1 vaccines: A randomized study in European adults. *Influenza and Other Respiratory Viruses*, *7*, 55–65.

Goto, N., & Akama, K. (1982). Histopathological studies of reactions in mice injected with aluminum-adsorbed tetanus toxoid. *Microbiology and Immunology*, *26*, 1121–1132.

Goto, N., Kato, H., Maeyama, J., Shibano, M., Saito, T., Yamaguchi, J., et al. (1997). Local tissue irritating effects and adjuvant activities of calcium phosphate and aluminium hydroxide with different physical properties. *Vaccine*, *15*, 1364–1371.

Graham, B. S., McElrath, M. J., Keefer, M. C., Rybczyk, K., Berger, D., Weinhold, K. J., et al. (2010). Immunization with cocktail of HIV-derived peptides in montanide ISA-51 is immunogenic, but causes sterile abscesses and unacceptable reactogenicity. *PLoS One*, *5*, e11995.

Greenwood, D. L., Dynon, K., Kalkanidis, M., Xiang, S., Plebanski, M., & Scheerlinck, J. P. (2008). Vaccination against foot-and-mouth disease virus using peptides conjugated to nano-beads. *Vaccine*, *26*, 2706–2713.

Grgacic, E. V., & Anderson, D. A. (2006). Virus-like particles: Passport to immune recognition. *Methods*, *40*, 60–65.

Gursel, I., Gursel, M., Ishii, K. J., & Klinman, D. M. (2001). Sterically stabilized cationic liposomes improve the uptake and immunostimulatory activity of CpG oligonucleotides. *The Journal of Immunology*, *167*, 3324–3328.

Harper, D. M. (2009). Currently approved prophylactic HPV vaccines. *Expert Review of Vaccines*, *8*, 1663–1679.

Harris, J., & Werling, D. (2003). Binding and entry of respiratory syncytial virus into host cells and initiation of the innate immune response. *Cellular Microbiology*, *5*, 671–680.

Harrison, W. T. (1935). Some observations on the use of alum precipitated diphtheria toxoid. *American Journal of Public Health and the Nation's Health*, *25*, 298–300.

Hartmann, G., Battiany, J., Poeck, H., Wagner, M., Kerkmann, M., Lubenow, N., et al. (2003). Rational design of new CpG oligonucleotides that combine B cell activation with high IFN-α induction in plasmacytoid dendritic cells. *European Journal of Immunology*, *33*, 1633–1641.

Hartmann, G., & Krieg, A. M. (2000). Mechanism and function of a newly identified CpG DNA motif in human primary B cells. *The Journal of Immunology*, *164*, 944–952.

Haynes, L. M., Jones, L. P., Barskey, A., Anderson, L. J., & Tripp, R. A. (2003). Enhanced disease and pulmonary eosinophilia associated with formalin-inactivated respiratory syncytial virus vaccination are linked to G glycoprotein CX3C-CX3CR1 interaction and expression of substance P. *Journal of Virology*, *77*, 9831–9844.

Heikenwalder, M., Polymenidou, M., Junt, T., Sigurdson, C., Wagner, H., Akira, S., et al. (2004). Lymphoid follicle destruction and immunosuppression after repeated CpG oligodeoxynucleotide administration. *Nature Medicine*, *10*, 187–192.

Hemmi, H., Takeuchi, O., Kawai, T., Kaisho, T., Sato, S., Sanjo, H., et al. (2000). A Toll-like receptor recognizes bacterial DNA. *Nature*, *408*, 740–745.

Herzog, C., Hartmann, K., Kunzi, V., Kürsteiner, O., Mischler, R., Lazar, H., et al. (2009). Eleven years of Inflexal V—A virosomal adjuvanted influenza vaccine. *Vaccine*, *27*, 4381–4387.

Hogenesch, H. (2013). Mechanism of immunopotentiation and safety of aluminum adjuvants. *Frontiers in Immunology*, *3*, 406.

Iwasaki, A., & Medzhitov, R. (2004). Toll-like receptor control of the adaptive immune responses. *Nature Immunology*, *5*, 987–995.

Jackson, D. C., Purcell, A. W., Fitzmaurice, C. J., Zeng, W., & Hart, D. N. (2002). The central role played by peptides in the immune response and the design of peptide-based vaccines against infectious diseases and cancer. *Current Drug Targets*, *3*, 175–196.

Karbach, J., Gnjatic, S., Bender, A., Neumann, A., Weidmann, E., Yuan, J., et al. (2010). Tumor-reactive CD8+ T-cell responses after vaccination with NY-ESO-1 peptide, CpG 7909 and Montanide ISA-51: Association with survival. *International Journal of Cancer, 126*, 909–918.

Kenter, G. G., Welters, M. J., Valentijn, A. R., Lowik, M. J., Berends-van der Meer, D. M., Vloon, A. P., et al. (2009). Vaccination against HPV-16 oncoproteins for vulvar intraepithelial neoplasia. *The New England Journal of Medicine, 361*, 1838–1847.

Kim, Y. M., Brinkmann, M. M., Paquet, M. E., & Ploegh, H. L. (2008). UNC93B1 delivers nucleotide-sensing toll-like receptors to endolysosomes. *Nature, 452*, 234–238.

Kim, D., Jung, J., Lee, Y., & Kwon, H. J. (2011). Novel immunostimulatory phosphodiester oligodeoxynucleotides with CpT sequences instead of CpG motifs. *Molecular Immunology, 48*, 1494–1504.

Kim, D., Kwon, S., Ahn, C. S., Lee, Y., Choi, S. Y., Park, J., et al. (2011). Adjuvant effect of liposome-encapsulated natural phosphodiester CpG-DNA. *BMB Reports, 44*, 758–763.

Kim, D., Kwon, H. J., & Lee, Y. (2011). Activation of Toll-like receptor 9 and production of epitope specific antibody by liposome-encapsulated CpG-DNA. *BMB Reports, 44*, 607–612.

Kim, D., Kwon, S., Rhee, J. W., Kim, K. D., Kim, Y. E., Park, C. S., et al. (2011). Production of antibodies with peptide-CpG-DNA-liposome complex without carriers. *BMC Immunology, 12*, 29.

Kim, Y. E., Kwon, S., Wu, G., Kim, D., Park, B. K., Park, J. A., et al. (2014). Therapeutic effect of a TM4SF5-specific monoclonal antibody against colon cancer in a mouse model. *Oncotarget, 5*, 8402–8415.

Kim, D., Lee, Y., & Kwon, H. J. (2013). Expression of IFN-γ induced by CpG-DNA stimulation in a human myeloid leukemia cell line KG-1. *Journal of the Korean Society for Applied Biological Chemistry, 56*, 553–558.

Kim, D., Rhee, J. W., Kwon, S., Kim, Y. E., Choi, S. Y., Park, J., et al. (2010). Enhancement of immunomodulatory activity by liposome-encapsulated natural phosphodiester bond CpG-DNA in a human B cell line. *BMB Reports, 43*, 250–256.

Kim, D., Rhee, J. W., Kwon, S., Sohn, W. J., Lee, Y., Kim, D. W., et al. (2009). Immunostimulation and anti-DNA antibody production by backbone modified CpG-DNA. *Biochemical and Biophysical Research Communications, 379*, 362–367.

Kong, X., San Juan, H., Kumar, M., Behera, A. K., Mohapatra, A., Hellermann, G. R., et al. (2003). Respiratory syncytial virus infection activates STAT signaling in human epithelial cells. *Biochemical and Biophysical Research Communications, 306*, 616–622.

Kool, M., Soullié, T., van Nimwegen, M., Willart, M. A., Muskens, F., Jung, S., et al. (2008). Alum adjuvant boosts adaptive immunity by inducing uric acid and activating inflammatory dendritic cells. *The Journal of Experimental Medicine, 205*, 869–882.

Krieg, A. M. (2002). CpG motifs in bacterial DNA and their immune effects. *Annual Review of Immunology, 20*, 709–760.

Krieg, A. M., Yi, A. K., Matson, S., Waldschmidt, T. J., Bishop, G. A., Teasdale, R., et al. (1995). CpG motifs in bacterial DNA trigger direct B cell activation. *Nature, 374*, 546–549.

Kwon, S., Choi, K. C., Kim, Y. E., Ha, Y. W., Kim, D., Park, B. K., et al. (2014). Monoclonal antibody targeting of the cell surface molecule TM4SF5 inhibits the growth of hepatocellular carcinoma. *Cancer Research, 74*, 3844–3856.

Kwon, H. J., & Kim, D. S. (2003). Regulation of macrophage inflammatory protein-2 gene expression in response to oligodeoxynucleotide containing CpG motifs in RAW 264.7 cells. *Biochemical and Biophysical Research Communications, 308*, 608–613.

Kwon, S., Kim, Y. E., Kim, D., Park, B. K., Wu, G., Kim, T. H., et al. (2013). Prophylactic effect of a peptide vaccine targeting TM4SF5 against colon cancer in a mouse model. *Biochemical and Biophysical Research Communications, 435*, 134–139.

Kwon, S., Kim, D., Park, B. K., Cho, S., Kim, K. D., Kim, Y. E., et al. (2012). Prevention and therapy of hepatocellular carcinoma by vaccination with TM4SF5 epitope-CpG-DNA-liposome complex without carriers. *PLoS One*, *7*, e33121.

Kwon, S., Kim, Y. E., Park, J. A., Kim, D. S., Kwon, H. J., & Lee, Y. (2014). Therapeutic effect of a TM4SF5-specific peptide vaccine against colon cancer in a mouse model. *BMB Reports*, *47*, 215–220.

Kwon, S., Kim, D., Park, B. K., Wu, G., Park, M. C., Ha, Y. W., et al. (2013). Induction of immunological memory response by vaccination with TM4SF5 epitope-CpG-DNA-liposome complex in mouse hepatocellular carcinoma model. *Oncology Reports*, *29*, 735–740.

Kwon, H. J., Lee, K. W., Yu, S. H., Han, J. H., & Kim, D. S. (2003). NF-κB-dependent regulation of tumor necrosis factor-alpha gene expression by CpG-oligodeoxynucleotides. *Biochemical and Biophysical Research Communications*, *311*, 129–138.

Lee, K. W., Jung, J., Lee, Y., Kim, T. Y., Choi, S. Y., Park, J., et al. (2006). Immunostimulatory oligodeoxynucleotide isolated from genome wide screening of *Mycobacterium bovis* chromosomal DNA. *Molecular Immunology*, *43*, 2107–2118.

Lee, S. A., Lee, S. Y., Cho, I. H., Oh, M. A., Kang, E. S., Kim, Y. B., et al. (2008). Tetraspanin TM4SF5 mediates loss of contact inhibition through epithelial-mesenchymal transition in human hepatocarcinoma. *Journal of Clinical Investigation*, *118*, 1354–1366.

Lee, K. W., Lee, Y., Kim, D. S., & Kwon, H. J. (2006). Direct role of NF-κB activation in Toll-like receptor-triggered HLA-DRA expression. *European Journal of Immunology*, *36*, 1254–1266.

Lee, K. W., Lee, Y., Kwon, H. J., & Kim, D. S. (2005). Sp1-associated activation of macrophage inflammatory protein-2 promoter by CpG-oligodeoxynucleotide and lipopolysaccharide. *Cellular and Molecular Life Sciences*, *62*, 188–198.

Liang, H., Nishioka, Y., Reich, C. F., Pisetsky, D. S., & Lipsky, P. E. (1996). Activation of human B cells by phosphorothioate oligodeoxynucleotides. *Journal of Clinical Investigation*, *98*, 1119–1129.

Magnusson, M., Tobes, R., Sancho, J., & Pareja, E. (2007). Natural DNA repetitive extragenic sequences from Gram-negative pathogens strongly stimulate TLR9. *The Journal of Immunology*, *179*, 31–35.

Malherbe, L., Mark, L., Fazilleau, N., McHeyzer-Williams, L. J., & McHeyzer-Williams, M. G. (2008). Vaccine adjuvants alter TCR-based selection thresholds. *Immunity*, *28*, 698–709.

Marshall, J. D., Fearon, K., Abbate, C., Subramanian, S., Yee, P., Gregorio, J., et al. (2003). Identification of a novel CpG DNA class and motif that optimally stimulate B cell and plasmacytoid dendritic cell functions. *Journal of Leukocyte Biology*, *73*, 781–792.

Mbow, M. L., De Gregorio, E., Valiante, N. M., & Rappuoli, R. (2010). New adjuvants for human vaccines. *Current Opinion in Immunology*, *22*, 411–416.

Mond, J. J., Lees, A., & Snapper, C. M. (1995). T cell-independent antigens type 2. *Annual Review of Immunology*, *13*, 655–692.

Morefield, G. L., Sokolovska, A., Jiang, D., HogenEsch, H., Robinson, J. P., & Hem, S. L. (2005). Role of aluminum-containing adjuvants in antigen internalization by dendritic cells in vitro. *Vaccine*, *23*, 1588–1595.

Muller-Pillasch, F., Wallrapp, C., Lacher, U., Friess, H., Buchler, M., Adler, G., et al. (1998). Identification of a new tumour-associated antigen TM4SF5 and its expression in human cancer. *Gene*, *208*, 25–30.

Mutwiri, G. K., Nichani, A. K., Babiuk, S., & Babiuk, L. A. (2004). Strategies for enhancing the immunostimulatory effects of CpG oligodeoxynucleotides. *Journal of Controlled Release*, *97*, 1–17.

Naz, R. K., & Dabir, P. (2007). Peptide vaccines against cancer, infectious diseases, and conception. *Frontiers in Bioscience*, *12*, 1833–1844.

Neuzil, K. M., Jackson, L. A., Nelson, J., Klimov, A., Cox, N., Bridges, C. B., et al. (2006). Immunogenicity and reactogenicity of 1 versus 2 doses of trivalent inactivated influenza vaccine in vaccine-naive 5-8-year-old children. *The Journal of Infectious Diseases*, *194*, 1032–1039.

Nohynek, H., Jokinen, J., Partinen, M., Vaarala, O., Kirjavainen, T., Sundman, J., et al. (2012). AS03 adjuvanted AH1N1 vaccine associated with an abrupt increase in the incidence of childhood narcolepsy in Finland. *PLoS One*, 7, e33536.

O'Hagan, D. T., Ott, G. S., Nest, G. V., Rappuoli, R., & Giudice, G. D. (2013). The history of MF59(®) adjuvant: A phoenix that arose from the ashes. *Expert Review of Vaccines*, *12*, 13–30.

Oussoren, C., Velinova, M., Scherphof, G., van der Want, J. J., van Rooijen, N., & Storm, G. (1998). Lymphatic uptake and biodistribution of liposomes after subcutaneous injection. IV. Fate of liposomes in regional lymph nodes. *Biochimica et Biophysica Acta*, *1370*, 259–272.

Park, M. C., Kim, D. B., Lee, Y., & Kwon, H. J. (2013). CD83 expression induced by CpG-DNA stimulation in a macrophage cell line RAW 264.7. *BMB Reports*, *46*, 448–453.

Park, B. K., Kim, D., Wu, G., Choi, S. H., Kim, D. S., Lee, Y., et al. (2014). Effect of epitope-CpG-DNA-liposome complex without carriers on vaccination of respiratory syncytial virus infection. *Journal of the Korean Society for Applied Biological Chemistry*, *57*, 667–676.

Peiris, J. S., de Jong, M. D., & Guan, Y. (2007). Avian influenza virus (H5N1): A threat to human health. *Clinical Microbiology Reviews*, *20*, 243–267.

Peiris, M., Yuen, K. Y., Leung, C. W., Chan, K. H., Ip, P. L., Lai, R. W., et al. (1999). Human infection with influenza H9N2. *Lancet*, *354*, 916–917.

Podda, A. (2001). The adjuvanted influenza vaccines with novel adjuvants: Experience with the MF59-adjuvanted vaccine. *Vaccine*, *19*, 2673–2680.

Purcell, A. W., Zeng, W., Mifsud, N. A., Ely, L. K., Macdonald, W. A., & Jackson, D. C. (2003). Dissecting the role of peptides in the immune response: Theory, practice and the application to vaccine design. *Journal of Peptide Science*, *9*, 255–281.

Rhee, J. W., Kim, D., Park, B. K., Kwon, S., Cho, S., Lee, I., et al. (2012). Immunization with a hemagglutinin-derived synthetic peptide formulated with a CpG-DNA-liposome complex induced protection against lethal influenza virus infection in mice. *PLoS One*, 7, e48750.

Riese, P., Schulze, K., Ebensen, T., Prochnow, B., & Guzmán, C. A. (2013). Vaccine adjuvants: Key tools for innovative vaccine design. *Current Topics in Medicinal Chemistry*, *13*, 2562–2580.

Samulowitz, U., Weber, M., Weeratna, R., Uhlmann, E., Noll, B., Krieg, A. M., et al. (2010). A novel class of immune-stimulatory CpG oligodeoxynucleotides unifies high potency in type I interferon induction with preferred structural properties. *Oligonucleotides*, *20*, 93–101.

Sangha, R., & Butts, C. (2007). L-BLP25: A peptide vaccine strategy in non small cell lung cancer. *Clinical Cancer Research*, *13*, s4652–s4654.

Schreiber, T. H., Raez, L., Rosenblatt, J. D., & Podack, E. R. (2010). Tumor immunogenicity and responsiveness to cancer vaccine therapy: The state of the art. *Seminars in Immunology*, *22*, 105–112.

Schwarz, T. F., Horacek, T., Knuf, M., Damman, H. G., Roman, F., Dramé, M., et al. (2009). Single dose vaccination with AS03-adjuvanted H5N1 vaccines in a randomized trial induces strong and broad immune responsiveness to booster vaccination in adults. *Vaccine*, *27*, 6284–6290.

Sparwasser, T., Hultner, L., Koch, E. S., Luz, A., Lipford, G. B., & Wagner, H. (1999). Immunostimulatory CpG-oligodeoxynucleotides cause extramedullary murine hemopoiesis. *The Journal of Immunology*, *162*, 2368–2374.

Stein, C. A., Subasinghe, C., Shinozuka, K., & Cohen, J. S. (1988). Physicochemical properties of phosphorothioate oligodeoxynucleotides. *Nucleic Acids Research*, *16*, 3209–3221.

Sulzer, B., & Perelson, A. S. (1997). Immunons revisited: Binding of multivalent antigens to B cells. *Molecular Immunology*, *34*, 63–74.

Tam, J. P., Clavijo, P., Lu, Y. A., Nussenzweig, V., Nussenzweig, R., & Zavala, F. (1990). Incorporation of T and B epitopes of the circumsporozoite protein in a chemically defined synthetic vaccine against malaria. *The Journal of Experimental Medicine*, *171*, 299–306.

Thoelen, S., Van Damme, P., Mathei, C., Leroux-Roels, G., Desombere, I., Safary, A., et al. (1998). Safety and immunogenicity of a hepatitis B vaccine formulated with a novel adjuvant system. *Vaccine*, *16*, 708–714.

Tong, J. C., Tan, T. W., & Ranganathan, S. (2007). Methods and protocols for prediction of immunogenic epitopes. *Briefings in Bioinformatics*, *8*, 96–108.

Verthelyi, D., Ishii, K. J., Gursel, M., Takeshita, F., & Klinman, D. M. (2001). Human peripheral blood cells differentially recognize and respond to two distinct CpG motifs. *The Journal of Immunology*, *166*, 2372–2377.

Vesikari, T., Groth, N., Karvonen, A., Borkowski, A., & Pellegrini, M. (2009). MF59-adjuvanted influenza vaccine (FLUAD) in children: Safety and immunogenicity following a second year seasonal vaccination. *Vaccine*, *27*, 6291–6295.

Welters, M. J., Bijker, M. S., van den Eeden, S. J., Franken, K. L., Melief, C. J., Offringa, R., et al. (2007). Multiple CD4 and CD8 T-cell activation parameters predict vaccine efficacy in vivo mediated by individual DC-activating agonists. *Vaccine*, *25*, 1379–1389.

Wille-Reece, U., Flynn, B. J., Lore, K., Koup, R. A., Miles, A. P., Saul, A., et al. (2006). Toll-like receptor agonists influence the magnitude and quality of memory T cell responses after prime-boost immunization in nonhuman primates. *The Journal of Experimental Medicine*, *203*, 1249–1258.

Wilson, I. A., Skehel, J. J., & Wiley, D. C. (1981). Structure of the haemagglutinin membrane glycoprotein of influenza virus at 3 Å resolution. *Nature*, *289*, 366–373.

Wu, Y., Ellis, R. D., Shaffer, D., Fontes, E., Malkin, E. M., Mahanty, S., et al. (2008). Phase 1 trial of malaria transmission blocking vaccine candidates Pfs25 and Pvs25 formulated with montanide ISA 51. *PLoS One*, *3*, e2636.

Yanasarn, N., Sloat, B. R., & Cui, Z. (2011). Negatively charged liposomes show potent adjuvant activity when simply admixed with protein antigens. *Molecular Pharmaceutics*, *8*, 1174–1185.

Yasuda, K., Yu, P., Kirschning, C. J., Schlatter, B., Schmitz, F., Heit, A., et al. (2005). Endosomal translocation of vertebrate DNA activates dendritic cells via TLR9-dependent and -independent pathways. *The Journal of Immunology*, *174*, 6129–6136.

Zhao, Q., Matson, S., Herrera, C. J., Fisher, E., Yu, H., & Krieg, A. M. (1993). Comparison of cellular binding and uptake of antisense phosphodiester, phosphorothioate, and mixed phosphorothioate and methylphosphonate oligonucleotides. *Antisense Research and Development*, *3*, 53–66.

Zhao, Q., Waldschmidt, T., Fisher, E., Herrera, C. J., & Krieg, A. M. (1994). Stage-specific oligonucleotide uptake in murine bone marrow B-cell precursors. *Blood*, *84*, 3660–3666.

CHAPTER FIVE

Chemical Platforms for Peptide Vaccine Constructs

Suhas Ramesh*, Prabhakar Cherkupally*, Thavendran Govender*, Hendrik G. Kruger*, Fernando Albericio*,†,‡,§,¶,||, Beatriz G. de la Torre*,†,1

*Catalysis and Peptide Research Unit, School of Health Sciences, University of KwaZulu-Natal, Durban, South Africa
†School of Chemistry, Yachay Tech, Yachay City of Knowledge, Urcuqui, Ecuador
‡School of Chemistry and Physics, University of KwaZulu-Natal, Durban, South Africa
§CIBER-BBN, Networking Centre on Bioengineering, Biomaterials and Nanomedicine, Barcelona Science Park, Barcelona, Spain
¶Department of Organic Chemistry, University of Barcelona, Barcelona, Spain
||Institute for Research in Biomedicine (IRB Barcelona), Barcelona, Spain
[1]Corresponding author: e-mail address: beatriz.garcia@upf.edu

Contents

Advances in Protein Chemistry and Structural Biology, Volume 99
ISSN 1876-1623
http://dx.doi.org/10.1016/bs.apcsb.2015.03.005

Abstract

Knowledge of the sequences and structures of proteins from pathogenic microorganisms has been put to great use in the field of protein chemistry for the development of peptide-based vaccines. These vaccine constructs include chemically tailored, shorter peptidic fragments that can induce high immunogenicity, thus shunning the allergenic and nonimmunogenic part of the antigens. Based on this concept, several different chemistries have been pursued to obtain novel platforms onto which antigenic epitopes can be tethered, with the aim to achieve a higher antibody response. In this regard, here we attempt to summarize the chemical strategies developed for the presentation of peptide epitopes.

1. INTRODUCTION

Nearly two-thirds of the world's population suffers from transmittable diseases, of which more than 17 million people die annually. The major threats to mankind are AIDS, tuberculosis, malaria, hepatitis, and emerging diseases. In this scenario, a realistic and reasonable vaccine development approach is pressing. This challenge can be met using protein chemistry.

Dating back more than 200 years, "vaccinology" has experienced several significant developments and progress in recent decades (Hilleman, 2000). From a historical perspective, the vaccine system was not credited to be of scientific importance; however, it has experienced huge developments in recent times. More than 40 diseases are currently controlled by vaccination protocols. With the ample information now available on the fundamental concepts of mammalian immunological systems and their mode of action, significant effort is being channeled into novel vaccines (Esser et al., 2003; Medzhitov & Janeway, 2000). In the past, vaccines were produced from inactivated organisms or crude components of the pathogen. Proteins of an immunogenic nature were used as the main antigens to induce antibody production; however, their application to therapeutics has not been fully exploited.

As a result of the poor performance of these traditional approaches, other alternatives emerged—these moving the focus away from the whole pathogen and toward the peptide epitope. Comprising short and specific fragments of a specific pathogen, peptide epitopes are attractive substitutes for traditional vaccines. These new vaccines include antigen(s), delivery systems, and an adjuvant, which together result in an enhanced immune response (Rueckert & Guzman, 2012). Antigens are less immunogenic when administered alone, and hence adjuvants are added to boost antibody response. The use of small peptides for antibody production envisages two limitations, namely that these peptides show poor immunogenicity and they are prone to enzymatic degradation. However, these peptide epitopes offer the following advantages: (i) they are not infectious agents and there is no risk of release during preparation, (ii) there is no reversion leading to virulence and no risk of genetic integration or recombination, (iii) harmful sequences can be excluded, (iv) immunogenicity can be modulated by structural fine-tuning, (v) they can be purified and characterized in a straightforward manner, (vi) they can be produced on a large scale at a reasonable cost, and (vii) they are easy to store and transport—contrasting with traditional vaccines, which require special precautions from their synthesis up to transport (Patarroyo & Patarroyo, 2008; Purcell, McCluskey, & Rossjohn, 2007; Skwarczynski & Toth, 2011). Because of these advantages, synthetic peptides have received great attention for vaccine design purposes.

Several promising peptide vaccines are in different stages of clinical studies (Fig. 1; Li, Joshi, Singhania, Ramsey, & Murthy, 2014).

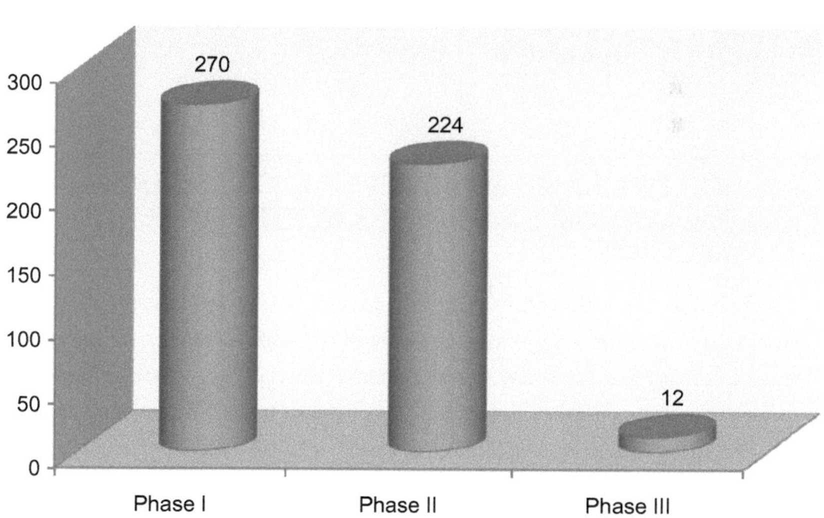

Figure 1 Bar graph showing the number of peptide-based vaccines that are in clinical trial stages. It is worth mentioning that extensive anticancer studies are under Phase III trials whereas other diseases, including viral diseases, tuberculosis, malaria, foot-and-mouth disease, among others, fall under Phase I and II trials.

An alternative to overcome the poor immunogenicity of small peptides is to conjugate them to carrier proteins like ovalbumin, bovine serum albumin, or keyhole limpet hemocyanin in order to trigger T cells. However, this option is also accompanied by several drawbacks, including the following: (a) the carrier itself could be immunogenic and could induce an unwanted immune response, (b) the arrangement of epitopes on the protein could be irregular, and (c) through the conjugation method chosen, the carrier could alter determinants of the antigen. These limitations have spurred researchers to develop novel antigen presentation methods that stimulate antibody response. Supplementation of the vaccines with adjuvants was an ideal approach, but only a few were successful since many exhibited severe side effects, and in some cases the adjuvants themselves were immunogenic (Rueckert & Guzman, 2012).

To date, only few methods have been added to the repository of peptide-based vaccines area. This could be attributed to the complexity of identifying specific epitopes and the difficulties encountered with their synthesis, purification, and characterization. However, those methods that have been developed have made significant contributions to this field. Most peptide-based vaccines are synthetically obtained by means of chemical strategies. To the best of our knowledge, there has been no attempt until now to compile the chemical platforms available for antigen presentation. This chapter seeks to provide a general overview of peptide epitope presentation strategies. Specifically, the next section describes chemical approaches for the preparation of peptide-based vaccines in order of their invention.

2. MULTIPLE ANTIGENIC PEPTIDE SYSTEM USING A PolyLys CORE: A BOON TO SYNTHETIC PEPTIDE VACCINES

Synthetic peptides can raise reactive antibodies with equivalent sequences as those in the original proteins (Lerner, 1982). Unsuccessful and unsatisfactory results from the earlier methods of vaccinology propelled researchers to design chemically unequivocal systems for generating antibodies in animals and vaccines for humans. To address this issue, in 1988 Tam developed a convenient and versatile approach called multiple antigen peptide (MAP) system to generate antipeptide antibodies, and therefore synthetic vaccines. This system comprises a peptidyl core matrix with radially branching synthetic peptides as dendritic arms (Fig. 2). The core matrix is formed by heptalysine with eight peptide arms of 9–16 amino acid residues.

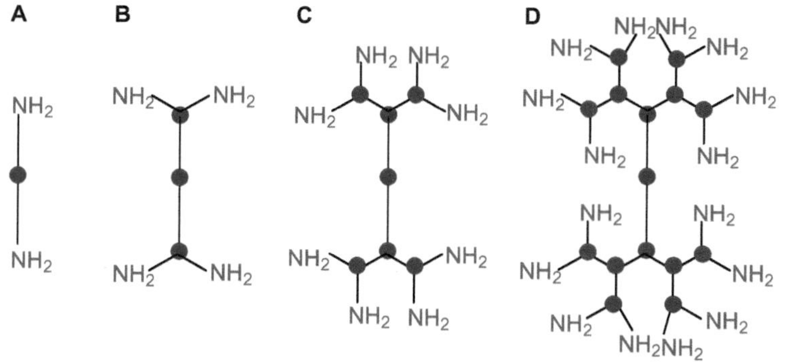

Figure 2 Schematic representation of MAPs containing a Lys core matrix with (**A**) first level, divalent; (**B**) second level, tetravalent; (**C**) third level, octavalent; and (**D**) fourth level, hexadecavalent; each amino function is tethered to a peptide to obtain a synthetic vaccine.

Solid-phase methodology was applied to synthesize a whole MAP with a molecular weight >10,000 which was directly used for immunization. In that study, six kinds of MAPs that elicited specific antibodies in rabbits, as well as in mice, were synthesized. Five of these reacted with their corresponding native proteins. This method has the advantage that it does not involve the conjugation of protein to a carrier.

The MAP comprised mainly three salient features: (a) an uncomplicated amino acid-like glycine or β-alanine, which, during the synthetic step, was attached to benzhydrylamine resin, which finally requires anhydrous hydrogen fluoride (HF) for global deprotection and detachment, (b) an inner core consisting of trifunctional groups possessing amino acid-like Lys, which enables branching (2–4 level), and (c) an acetylated peptide tethered to the inner core matrix, which lies on the outer surface. Boc-Lys(Boc) was used as both N^{α} and N^{ε}-amino groups were available as reactive ends. Sequential propagation of Boc-Lys(Boc) generated 2^n reactive ends. In this study, a MAP with an octa-branching core matrix consisting of a heptalysine was synthesized as a prototype. Peptide antigens had Boc groups for α-amino-termini protection and benzyl alcohol derivatives for most side-chain amino acids. The crude MAP-branched peptide oligolysine was obtained after cleavage from the resin by low–high HF method or the low–high trifluoromethane-sulfonic acid method. After thorough dialysis, the MAP was lyophilized and purified by high-performance gel permeation or ion exchange chromatography.

One of the characteristic features of MAPs is that they have smaller inner core matrix and high density of homogeneous peptidic antigens. All the

MAPs synthesized in this study provided high-titer antibodies, thus proving that this method can be conveniently used for the production of vaccines. However, the authors drew attention to a number of issues to be resolved, such as molecular arrangement, size, shape, flexibility, and level of branching.

3. SYNTHETIC ANTIGEN CARRIER BEARING *N*-(*S*-ACETYLMERCAPTOACETYL)-GLUTAMYL RESIDUES

Drijfhout and Bloemhoff (1991) described a novel synthetically functionalized antigen carrier that consisted of seven branched Lys residues in which each Lys consisted of two *N*-(*S*-acetylmercaptoacetyl)-glutamyl residues at its N-terminus. This method was devised in view of some of the limitations posed by Tam's MAP system. The original MAP approach is not suitable for the preparation of oligosaccharides or glycopeptides-based antigens because of the harsh deprotection and cleavage conditions used. Moreover, during the synchronized synthesis of eight antigenic peptides, steric hindrance would lead to impurities. Furthermore, chemical analysis and purification would be cumbersome. In contrast, the new synthetic approach by this research group did not apply harsh deprotection conditions and, importantly, separately synthesized antigen was conjugated to a previously prepared totally functionalized lysine core, which can be kept in stock (Fig. 3).

Fmoc-4-aminobutyric acid (Abu) was attached to Wang resin and the lysine tree was built by three successive couplings of Fmoc-Lys(Fmoc)-OPfp. After Fmoc removal from the α- and ε-ends, Fmoc–Glu(*t*Bu)–OPfp was coupled, and the chain was elongated with eight *S*-acetylmercaptoacetyl groups. Functionalized peptide carrier was obtained by removing *t*Bu groups with aq. TFA, which had acceptable water solubility. The *S*-acetyl function of the carrier was removed using hydroxyl amine at pH 8.0 and reacted with excess of *S*-Npys-cysteinyl antigenic peptide (Npys: 3-nitro-2-pyridinesulfenyl)—a sequence that matches the alignment of amino acids 9–21 of herpes simplex virus type 1 glycoprotein D. The course of the reaction was monitored by UV–vis spectrometry, which measured disulfide bond formation and the release of 3-nitro-2-thiopyridone. Not only can peptide antigens be stapled but also glycopeptides and saccharide-derived antigens. This system resulted in a high molecular weight conjugate with remarkable high antigen content.

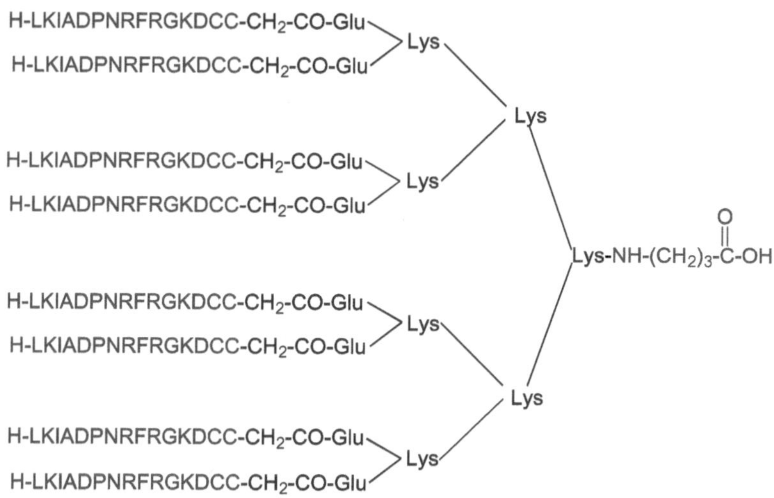

Figure 3 Representation of a MAP obtained by the reaction of acetyl-protected antigen carrier (blue/green (dark/light gray in the print version)) and Cys-containing antigenic peptide (red (gray in the print version)); green (light gray in the print version) represents aminobutyric acid, which acts as a reference amino acid, and "I" in the antigenic peptide is Nle.

4. THIOESTER AS A REPLACEMENT FOR NATURAL PEPTIDE BOND (CHEMICAL LIGATION): CONSTRUCTION OF PROTEINS

As these peptide vaccine constructs can be considered proteins on the basis of their size, the use of protein engineering techniques is feasible for their synthesis. Although site-directed mutagenesis (Smith, 1985), genetic manipulation (Wells, Vasser, & Powers, 1985), and other (Noren, Anthony-Cahill, Griffith, & Schultz, 1989) methods have been successfully applied for the synthesis of proteins, the chemical approach (Kent, 1988) offers more scope and greater flexibility to incorporate noncoded moieties, ranging from unnatural amino acids (Wlodawer et al., 1989) to fixed elements of 3D structure. In an ideal approach, protein engineering consists of chemically ligating readily available synthetic peptides without protections to yield a polypeptide with a number of modification(s). Followed by Schnoelzer and Kent (1992), this method allowed them to use extremely selective chemistry for the ligation of reactants with tailored functionalities. This approach offers potential advantages, for example, other functional groups in the reactants do not need to be associated with protecting groups.

Furthermore, it provides flexibility and a general method for the synthesis of other protein moieties.

Although this strategy was not applied directly to vaccine preparation, we consider it pertinent to discuss this dovetailing method as it could be adopted for the construction of peptide-based vaccines. The authors chose HIV-1 protease as the synthetic target, in which the unprotected N-terminal half was reacted with the corresponding unprotected C-terminal half with a bridging pseudopeptide linkage (Fig. 4). High selectivity was achieved when both halves of the polypeptide have exclusive reactive functional groups at the N- and C-terminals. Here, the nucleophilic thiol group on one chain was used to attack an alkyl bromide on the other segment, thereby enabling covalent linkage between two halves by means of a thioester bond.

The preparation of proteins by selective chemical ligation offers a two-fold advantage, namely efficient synthesis and simultaneous introduction of a new motif into the polypeptide/protein. This approach also paves the way for the successful syntheses of a number of fascinating and significant modifications of protein structure, including the alterations of stereochemistry by the introduction of D-amino acids, the inclusion of fixed elements of 3D structure, and the hybridization of protein–nonprotein macromolecules.

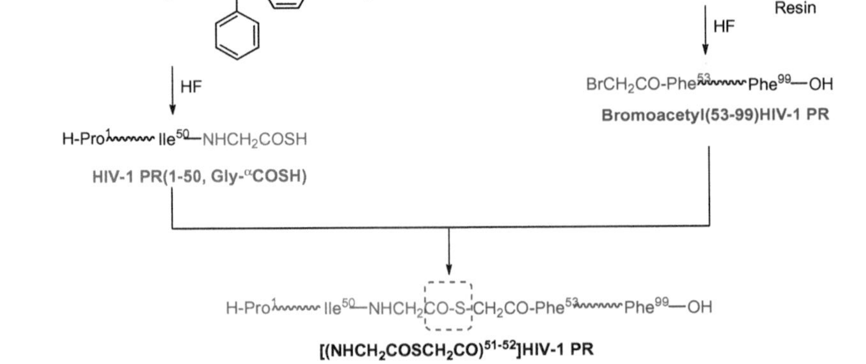

Figure 4 Sketch of dovetailing unprotected synthetic peptides through a thioester bond (boxed in red (gray in the print version)).

5. SYNTHETIC PEPTIDE DENDRIMER WITH THIAZOLIDINYL PENDANT

Engineered artificial proteins that act as enzymes (Hahn, Klis, & Stewart, 1990), ion channels (Montal, Montal, & Tomich, 1990), diagnostic reagents (Tam & Zavala, 1989), and vaccines (Tam, 1988) came into being because of the flexible nature of peptide dendrimers—a feature that has facilitated the use of these structures in a number of applications. Although dendritic peptides offer many opportunities, their synthesis is challenging. In this regard, their synthesis by SPPS results in unsatisfactory macromolecular products with low purity. The use of protected peptide monomers could be an appropriate choice to circumvent this problem, but these structures show less solubility and require more time for coupling. A more user friendly and proficient strategy is to have nonpeptidyl linkages to achieve chemical ligation between the matrix and unprotected peptides. Thioalkylation, thioester formation, and oximation are excellent examples of linkages for this approach.

A chemoselective reaction between an alkyl aldehyde and a weak base such as 1,2-amino thiol under acidic condition results in a thiazolidine ring. To demonstrate the feasibility of this reaction, Rao and Tam (1994) used an octavalent Lys-matrix (MAP) to construct a peptide dendrimer (Fig. 5). The N-terminal of the matrix consisted of Ser, which was oxidized using sodium periodate at pH 7 to obtain a glyoxylyl derivative of $(HCO)_8$-MAP. This

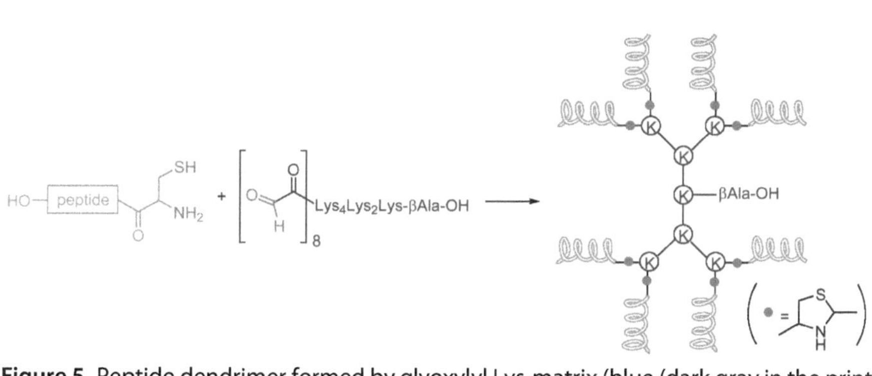

Figure 5 Peptide dendrimer formed by glyoxylyl Lys-matrix (blue (dark gray in the print version)) and an unprotected 24 amino acid-containing peptide (orange (light gray in the print version)) derived from the V3-loop of gp120, HIV-1 via thiazolidine linkage (pink (gray in the print version) dot). A new asymmetric carbon is generated in the thiazolidine ring.

matrix was reacted with a purified and unprotected peptide with Cys at the N-terminus. The authors found that the use of an organic cosolvent (NMP-H_2O) at high temperature (37 °C) gave good results. The formation of an octameric product took around 7 h, and various peptide dendrimers (tetra/penta/hexa/hepta) were observed at different time intervals. Purification and characterization were trouble-free.

This work was of great significance for several reasons: (i) it was the first attempt to achieve ligation *via* thiazolidine, (ii) good conformational rigidity was obtained, (iii) both the reactants were derived from readily available amino acids, and (iv) the peptide dendrimer obtained had a molecular weight of 24,000—the first of its kind.

6. OXIME CHEMISTRY FOR SYNTHETIC PEPTIDE-BASED VACCINES

In 1994, Rose used oxime chemistry to prepare artificial proteins. Later, his group applied the same technology to construct a tetrameric oxime vaccine based on peptides (Rose, Zeng, Brown, & Jackson, 1995) derived from influenza virus hemagglutinin (Fig. 6) and tested its activity in a model system.

The template (red (light gray in the print version)), which is in the Ser form, was subjected to sodium metaperiodate oxidation and later isolated as

Figure 6 Covalent structure of tetraoxime (green (gray in the print version)) containing four arms of influenza hemagglutinin peptide (purple (gray in the print version)) tethered to a cyclic template (red (light gray in the print version)).

Ac-CK(HCOCO)AK(HCOCO)PGK(HCOCO)AK(HCOCO)C-NH_2. This was followed by oximation of the oxidized template (red (light gray in the print version)) with amino-oxyacetyl peptide (purple (gray in the print version)) in acetate buffer of pH 4.6 for 16 h at rt. Analytical HPLC data revealed that tetraoxime was homogeneous, and ESI-MS revealed the expected mass value. The polyoxime approach was developed in view of the drawbacks of the previously established methods, as briefly explained below. First, a thioester linkage is superior chemically but the end product was found to be unstable at neutral pH and had a hydrolysis half-time of just 2 h. Second, the thioether bond-containing polypeptide was highly stable but was not resistant to alkylation, thus yielding several by-products. Third, domain ligation resulted in a peptide-linked product, but the rate of the reaction was slow. Finally, native chemical ligation also gave an amide-bonded product; however, this product has not yet adopted for testing polyimmunogens.

There is a prominent difference in homogeneity between polyoxime and other MAPs. In the latter chemistry, all the chains are simultaneously built with overcrowded amide bonds, and both the synthesis and purification is complex. In contrast, the polyoximation technique involves straightforward synthetic and purification steps that allow a clean reaction that leads to a homogeneous end product. Another notable feature of the oxime bond approach is the possibility to conjugate candidate epitopes through the N- and also C-terminal. The tetraoxime construct raises high antibody titers and also induces high levels of proliferating T cells *in vivo*.

7. POLYOXYETHYLENE-BASED HETEROBIFUNCTIONAL CROSS-LINKING REAGENTS

Synthetic peptides can act as epitopes in the preparation of vaccines. Given their small size and poor immunogenicity, these peptides need to be attached to large carriers, such as proteins, synthetic polymers, and liposomes. The attachment of short peptides to preforme liposomes was established by Martin and Papahadjopoulos (1982); however, this method had some drawbacks as the final product was intrinsically immunogenic and the spacer used showed poor hydrophobicity. In this context, Frisch, Boeckler, and Schuber (1996) synthesized a heterobifunctional reagent that carried polyoxyethylene spacers. Such spacers are more hydrophobic and less immunogenic. Furthermore, their length can be tailored to requirements. Three new heterobifunctional reagents were synthesized, one end

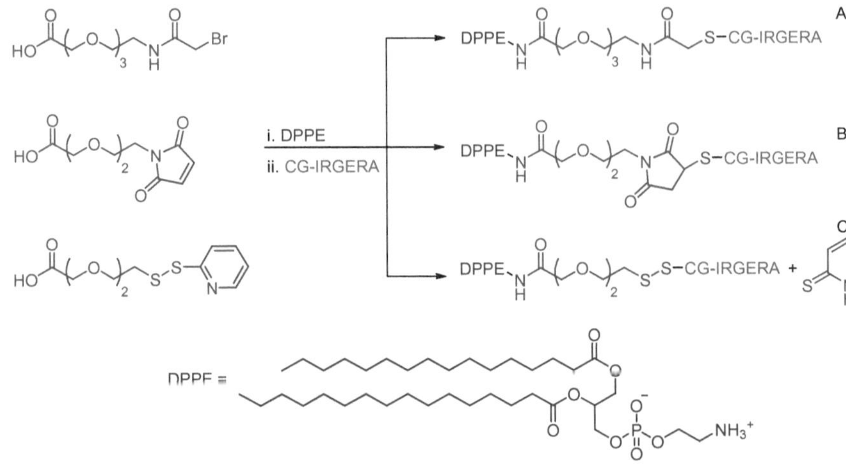

Figure 7 Short polyoxyethylene-based thiol-reactive heterobifunctional cross-linking reagents containing the following: (**A**) bromoacetyl (red (light gray in the print version)), (**B**) maleimide (green (light gray in the print version)), and (**C**) dithiopyridine (purple (gray in the print version)) moieties.

of which was functionalized to react with a thiol group of a peptide (CG-IRGERA) to give a stable thioether bond and the other end had a carboxyl function to ensure reaction with the amine of DPPE (Fig. 7).

A comparison of these moieties revealed that the order of reactivity was maleimide > dithiopyridine > bromoacetyl. These reagents also have applications in the fields of glycoconjugates, gene delivery, and allied systems.

8. DOUBLE ORTHOGONAL CHEMOSELECTIVE LIGATION VIA A HYDRAZONE LINKAGE

Dendritic cells (DCs) initiate and modulate the immune response (Banchereau & Steinman, 1998). They stimulate T-lymphocyte response by presenting antigens bound to major histocompatibility complex (MHC) class II molecules. Mannosylation of a peptide or protein has been reported to enhance the ability to stimulate peptide-specific T-cell clones 200- to 10,000-fold compared to non-mannosylated forms (Steinman, Pack, & Inaba, 1997). Therefore, the mannosylation of antigens increases their presentation by DCs, thereby improving the efficiency of vaccines and allowing their use at a lower dosage. In this regard, Grandjean, Rommens, Gras-Masse, and Melnyk (2000) developed a convergent approach that permitted the reaction of glycosides, antigens, and

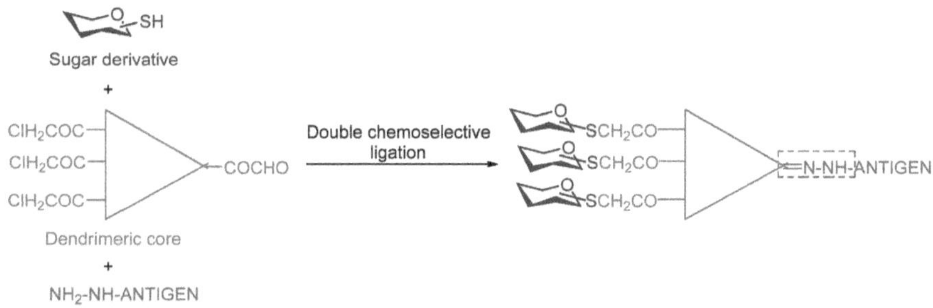

Figure 8 General schematic for double ligation strategy to antigen-tethered glycodendrimers; amine in the antigen (formed by hydrazinolysis of antigen) reacts with the aldehyde group of the dendritic core to form a hydrazone linkage (boxed); on the other hand, thiol-functionalized mannoside reacts with the chloroacetyl group of the dendritic core to form a thioether bond.

dendrimeric cores by double chemoselective ligation, achieving satisfactory yield and purity (Fig. 8). This method also expanded the number of synthetic routes for the construction and rapid optimization of antigen presentation.

This approach can be exemplified on the following grounds: (i) dendritic cores with two functionalities are chemically compatible, (ii) mannosides and antigens are functionalized with complementary functional groups, and (iii) one-pot synthesis is achieved by linking prepurified, fully deprotected reactants using two consecutive orthogonal chemoselective ligation reactions. The reaction of hydrazine antigen and the aldehydic dendritic core was carried out using DMF/citrate phosphate buffer (pH 5.2) to render clean hydrazone fragment. Next, the chloroacetylated derivative was treated with the mannosyl moiety at pH 8–8.5.

9. MULTIPLE ANTIGENIC PEPTIDES WITH SYNTHETIC CHIMERIC EPITOPES

Viral hepatitis shows greater prevalence in dialysis patients than in the general population. Also, those undergoing hemodialysis have great risk of infection by GB virus (GBV) (formerly called hepatitis G virus) strain C. In an effort to find a peptide-based vaccine for this infection, Gomara, Fernandez, Perez, Ercilla, and Haro (2010) devised a MAP in which chimeric (two different kinds) multiple antigens were used to construct a new system. They also compared the antigenic response of these new peptides with that of linear ones. This system has been tested against GBV, and various epitopes from E2, NS4 and NS5 GBV-C proteins were selected.

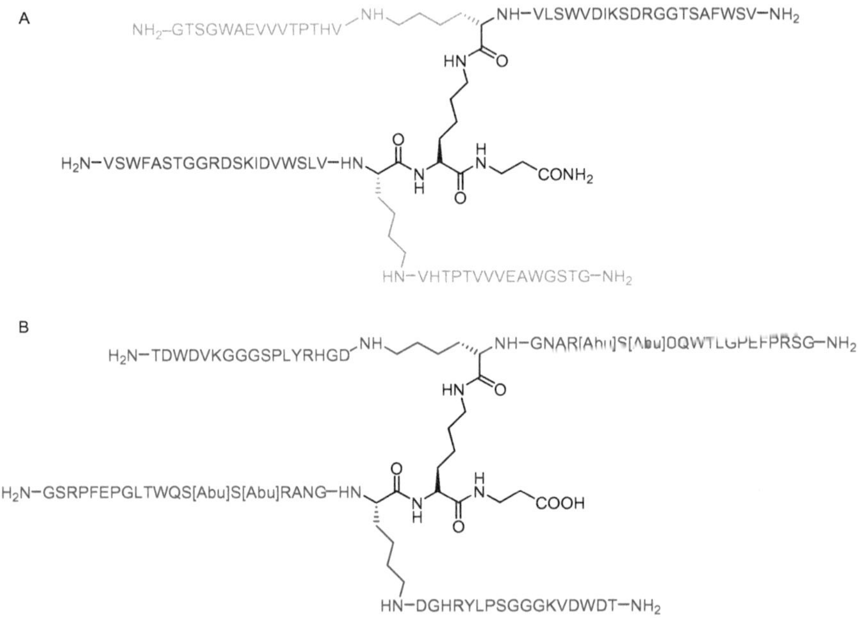

Figure 9 Structures of (**A**) MAP$_4$[E2(99–118)-NS5a(112–126)] and (**B**) MAP$_4$[E2(7–26)-NS4a(27–43)]; Abu: L-α-aminobutyric acid.

Two systems, namely MAP$_4$(E2-NS5a) and MAP$_4$(E2-NS4a) (Fig. 9), were synthesized in which Lys had Fmoc and Dde as orthogonal protections at ε-positions. Cys residues of E2(99–118) at 112/114 positions were replaced by 2-Abu, which has similar steric and hydrophobic properties (Haro, Perez, Garcia, Chan, & Ercilla, 2003). This replacement was to prevent Cys from undergoing dimerization.

The synergistic effect of presenting two copies of antigens from distinct GBV-C proteins in the same molecule had a pronounced effect on the antibody reactivity compared to monomeric linear forms, which had a weaker antigenic effect. Results of immunoassays demonstrated the potential of synthetic tetrameric branched peptides with surface and nonstructural GBV-C protein sequences as diagnostic tools hepatitis G.

10. GOLD NANOPARTICLES AS ANTIGEN CARRIERS

Recent approaches on developing peptide-based vaccines have focused mainly on the following two routes: (i) attachment of functional moieties like T-cell epitopes, cell-penetrating peptides, and lipo fatty acids;

and (ii) use of nanomaterials with the aid of self-assembling peptides, nonpeptidic dendrimers, and Au nanoparticles for antigen presentation. This subheading gives an insight into the application of nanoparticles, in particular those of Au for the preparation of synthetic vaccines. Apart from organic components like polymers and peptides, inorganic substances like SiO_2, Au, Ag, and Pt also form nanoparticles. In this regard, gold nanoparticles (GNPs) are of particular interest for biological applications, as well as for multimeric epitope presentation. These nanoparticles are popular mainly because they are biocompatible and nonimmunogenic. Furthermore, they can be easily prepared from the salt ($H[AuCl_4]$), whose particle size can be controlled. Thus, the surface of the Au electrodes can be easily reacted with thiol-bearing molecules (Fig. 10). The strong linkage between the thiol group and the gold surface enables the nanoparticles to be subjected to various chemoselective ligation reactions. GNPs deliver the target materials into the cell by endocytosis and they are resistant to enzymatic and lysosomal attack, and hence are ideal candidates for antigen presentation.

In a study by Chen, Hung, Lin, and Huang (2010) on foot-and-mouth disease vaccines, the addition of an extra Cys residue to the C-terminus of an antigenic peptide of VP1 protein permitted its conjugation to GNPs of different size and with high affinity. Those authors also described that the immunogenicity of the synthesized conjugates was dependent on size of the molecule, which also showed varied biodistribution. It is of interest to note that antigens bearing GNPs showed three times more antibody response than nonconjugated ones.

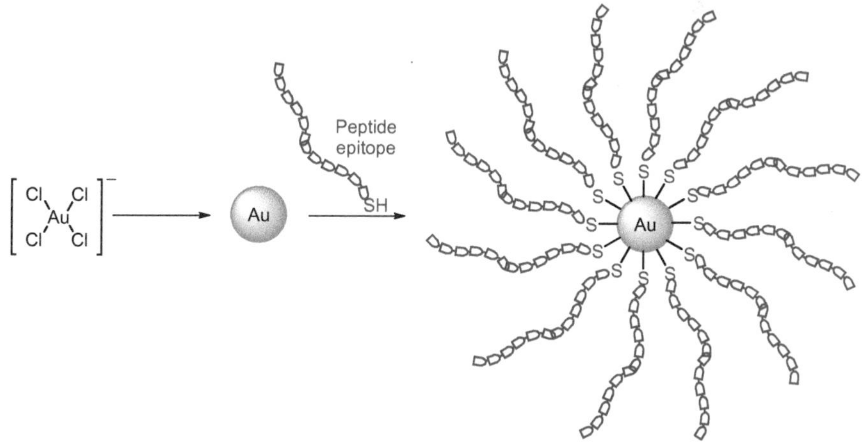

Figure 10 A general schematic of a multiple antigenic-presenting gold nanoparticle vaccine.

11. BRANCHED LIPOPEPTIDE CANDIDATE VACCINE

Given the urgent need to develop novel methodologies to improve the immunogenicity of peptides, lipidation emerges as a simple and highly reliable method to increase the biological half-life of these molecules (Brumeanu, Casares, Bot, Bot, & Bona, 1997). Lipidation also provides an adjuvant-like effect (Jackson et al., 2004) and has also been found to be more effective in cytosolic uptake and prolonged presentation events (Langhans et al., 2004). With this information in hand, Samayoa, Diaz-Mitoma, and Azizi (2011) constructed lipidated MAPs and compared their immunogenic response with that of other constructs, as depicted in Fig. 11; i.e., effect of lipidated peptide, branched peptide, or the combination of both on the immunogenic response. Known peptide antigen of hemaglutinin protein of A/Puerto Rico/8/34 was selected as model for this study.

The peptide candidates constructed had a simple system comprising (**A**) a T-helper epitope joined by two Gly spacers to a B-cell epitope, (**B**) lipidated peptide with plamitic acid at the N-terminus, and (**C**) single- and double Lys-branched peptides. Each of the three peptide structures was constructed using B-cell epitope as a backbone and T-helper epitope as branches and

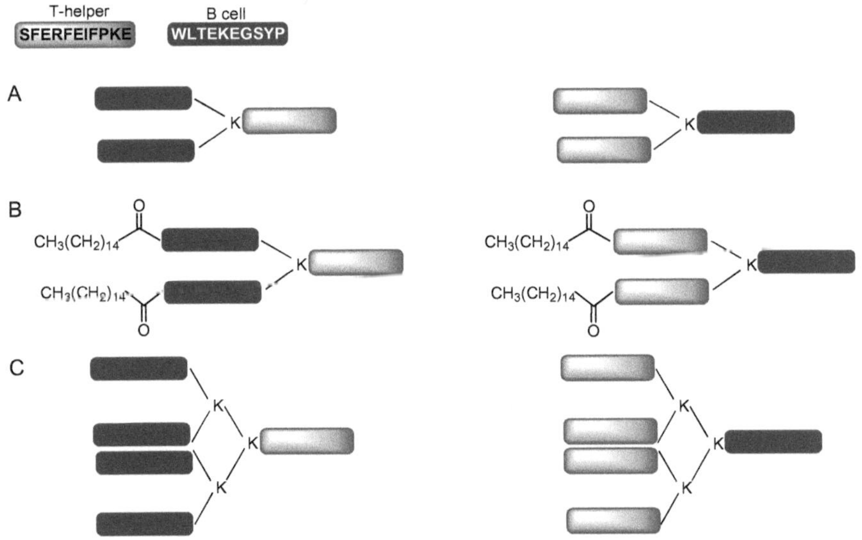

Figure 11 Schematic depiction of MAPs: (**A**) MAP with single-branched Lys, (**B**) lipidated MAP, and (**C**) double Lys-branched MAP.

vice versa. Results of antibody binding assays revealed that lipidated MAP, that is to say a combination of branched design and attachment of lipid moiety, had a high immune response, thus suggesting that peptide lipidation is a valuable strategy for vaccine development.

12. ANTIGENIC MIMICKING WITH CYS-BASED CYCLIZED PEPTIDES

The development of and antigenic-based vaccines for the treatment of classical swine fever virus (CSFV) is considered vital as this is a highly contagious disease that causes significant economic losses. The outer envelope of the virus contains glycoprotein E2, an immunodominant antigen that induces neutralizing antibodies during CSFV infection (Weiland et al., 1990). In a study carried out by Chang et al. (2012) the antigenic determinants on the B and C domains of this virus were addressed and synthetic cyclized peptides spanning these domains were designed in such a way as to mimic the original conformational epitopes. The authors used Chemically Linked Peptides on Scaffolds technology to cyclize the peptides (Timmerman, Puijk, & Meloen, 2007). In this approach, a linear peptide was chemically linked to a synthetic scaffold with conformational fixation of the peptide. Overlapped looped peptides with 20-mer/33-mer by one/two cyclized dicysteine bridges were synthesized (Fig. 12). The 20-mer dicysteine-containing and

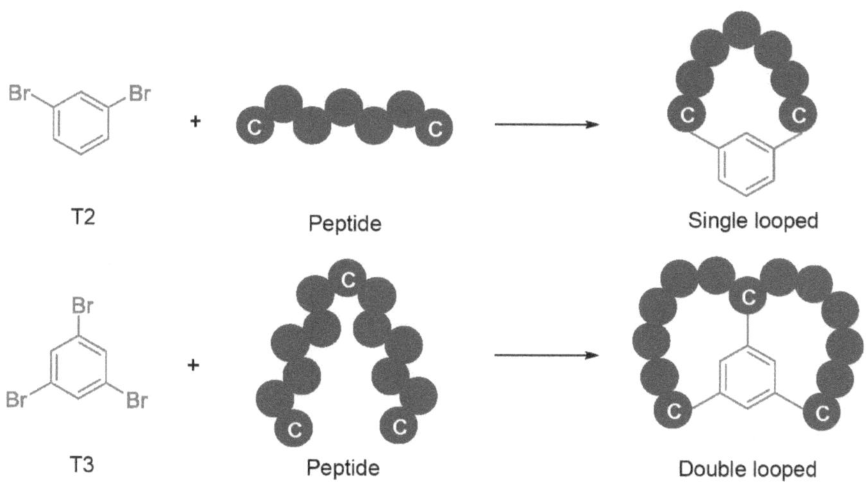

Figure 12 Cyclized single- and double-looped peptides.

the 33-mer tricysteine-containing overlapping peptides were assembled as the templates C-XXXXXXXXXXXXXXXXX-C-plate, treated with 1,3-bis(bromomethyl)benzene a,a-dibromoxylene (T2), and C-XXXXXXXXXXXXXXX-C-XXXXXXXXXXXXXXX-C-plate treated with 2,4,6-tris (bromomethyl)mesitylene (T3), to form single- and double-looped platforms, respectively.

Reactivity with antibodies showed that double-looped peptides were more powerful compared with single-looped ones. These cyclic peptides were specifically recognized by monoclonal antibodies, thus suggesting that their antigenicity was highly specific and analogous to the viral protein. Moreover, the vital amino acid residues responsible for antibody binding on cyclized peptides were identical to those on the expressed mutated proteins. Overall, these results reveal that peptide cyclization via dicysteine bridges and covalent linkages provide a system that mimics the B/C domains of the native epitope. Hence, the cyclized peptide may provide a suitable platform to expand the vaccine preparation for CSFV.

13. "REVERSE" THIOETHER LIGATION AS AN ALTERNATIVE ROUTE FOR MAPs CONSTRUCTION

Although classical thioether ligation proved a simple technique and yielded an acceptable number of targets that were metabolically stable, it has a serious limitation, namely that at physiological pH, which is necessary for thioether linkage, dimerization of the thiol peptide dominates over thioether formation. The only way to overcome this drawback is to use an excess of thiol peptide, which would also be wasted by means of dimerization. In order to avoid such waste and improve the reaction, Monso, Kowalczyk, Andreu, and de la Torre (2012) explored the possibility of conducting the thioether reaction in "reverse" mode. In the standard approach, the Lys core with haloacetyl functions is reacted with thiol peptide. In contrast, in the "reverse" mode, the Lys core is derivatized with thiol functions, and peptide epitopes as halogenated versions that result in the target dendrimer (Fig. 13).

Functionalization of the Lys core with thiol groups was done on solid-phase mediated by 1,3-diisopropylcarbodiimide (DIC)/3-tritylmercaptopropionic acid (Trt-Mpa). The advantages of using Trt-Mpa are that it is odorless and easy to handle. Moreover, the Trt protecting group is straightforward. Tris(2-carboxyethyl)phosphine was used in the reaction, which *in situ* reduces the disulfide formation. In the reverse

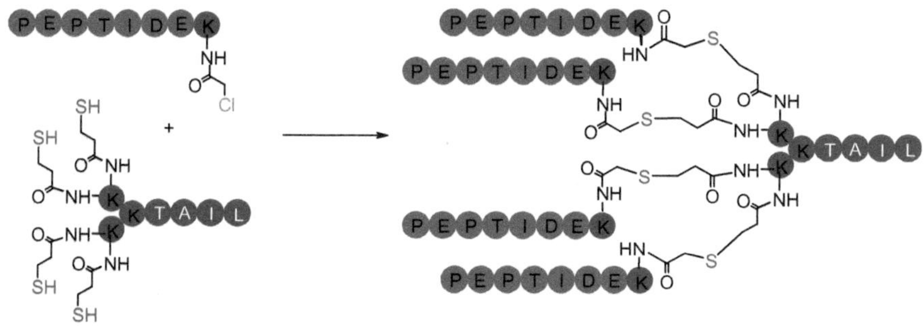

Figure 13 Reverse thioether-based approach for the synthesis of multimeric peptides.

technique, dimerization was not an issue, and also a more homogeneous product was formed, thus enabling easy purification and characterization. Another noteworthy feature is that just 2 equiv of haloacetyl peptide is sufficient per thiol group. Furthermore, the authors explored this reaction in solid-phase using ChemMatrix® resin. 3,3′-Dithiopropionic acid was used to functionalize the Lys-dendron as thiol, which was reacted with haloacetyl peptide to obtain a dendrimer. Though homogeneity was an issue, the authors claimed that this was the first record of on-resin ligation of unprotected peptides in aqueous medium, thus paving the way for other chemoselective reactions.

14. SELF-ADJUVANTING POLYMER–PEPTIDE CONJUGATES AS THERAPEUTIC VACCINE CANDIDATES

Dendrimers are structurally distinct artificial polymers with sizes and physicochemical properties as large molecules like proteins (Boas & Heegaard, 2004). Dendrimers can therefore be regarded as potential systems for peptide-based vaccine delivery since antigen molecules can be attached to their peripheries. Human papillomavirus (HPV) causes cervical cancer, a condition in which HPV proteins interfere with the cell cycle and the host cell. To date, the administration of a peptide-based therapeutic HPV vaccine has not been achieved without the aid of an adjuvant (Liu, Hussein, Toth, & Skwarczynski, 2012). Polyacrylate has been found to be a suitable candidate as amphiphilic dendrimer, proving to be a self-adjuvanting prophylactic vaccine (Skwarczynski et al., 2010; Zaman et al., 2011). Based on this observation, Liu et al. (2013) hypothesized that star polymer-based delivery platforms would have immunomodulating potency to develop T-cell mediated immunity. In this regard, they prepared polymer–peptide conjugates with HPV-16 E7 protein epitope. To synthesize the candidate vaccine against HPV-related cancers, a Cu-catalyzed alkyne-azide 1,3-dipolar cycloaddition "click reaction" was used (Fig. 14).

The efficacy of the synthesized conjugates demonstrated that these are self-adjuvanting systems with a therapeutic effect on tumors. This system exhibited strong immunogenic properties, thus eliminating the common toxic side effects that occur with external adjuvants. Furthermore, the

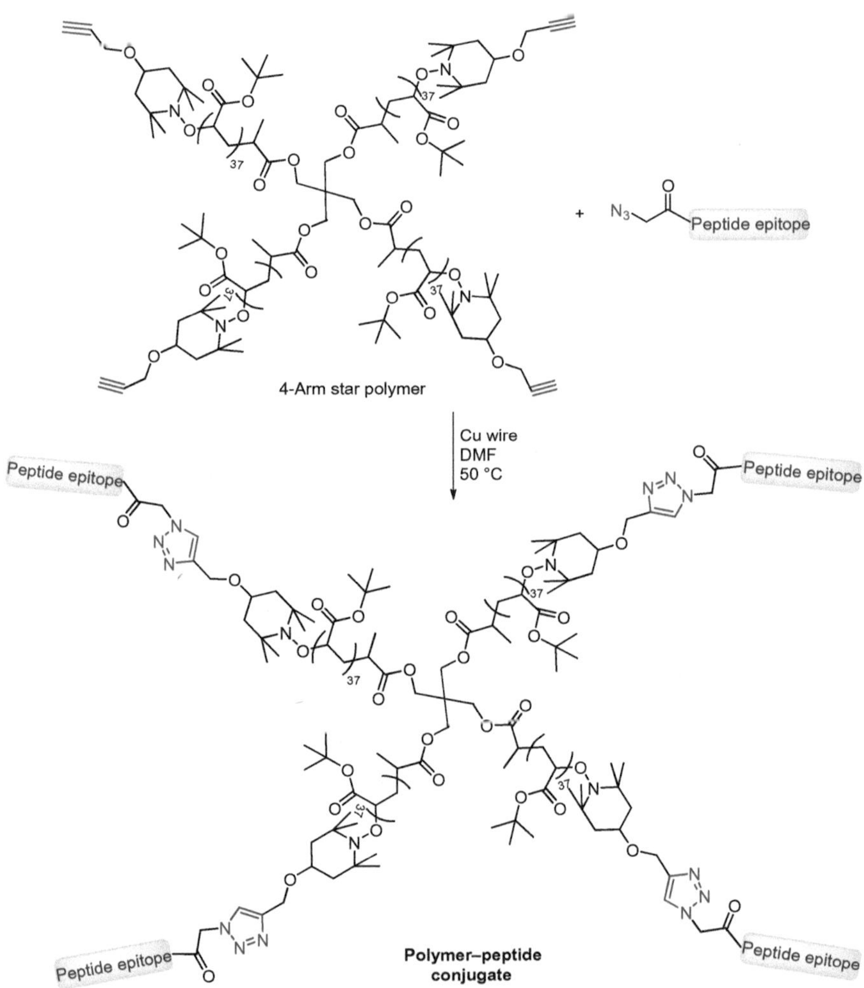

Figure 14 Synthesis of polymer–peptide conjugate by click chemistry between alkyne and azide to form triazole as the pendant moiety.

peptide epitope synthesis was greatly enhanced by just changing the regular SPPS and applying the isopeptide method.

15. MALEIMIDE AS A TOOL TO INCREASE PEPTIDE IMMUNOGENICITY

The preceding section gave details about the advantages/disadvantages of standard and "reverse" thioether chemistries; the latter emerging as the most promising. However, the availability of additional methods for the construction of peptide-based vaccines is always beneficial. In this regard, Monso, de la Torre, Blanco, Moreno, and Andreu (2013) explored an additional system for epitope linking through thiol-maleimide coupling. It is an example of nucleophilic conjugate addition to α,β-unsaturated carbonyls. This chemistry has also been used in peptide, DNA, and protein derivatization fields (Houseman, Gawalt, & Mrksich, 2003; Roberts, Bentley, & Harris, 2002). The treatment of the branched Lys core at both α- and ε-amino positions with *N*-maleoyl-β-alanine/DIC resulted in functionalized maleimide. Next, a twofold molar, stoichiometric amount of prepurified thiol peptide at pH 6.0 was added at rt. to obtain epitope-maleimido-Lys core peptide (Fig. 15), the reaction of which was completed within a minute.

The thiol-ene chemistry described herein used only one peptide equivalent per branch and produced minimal by-product formation, thus revealing that the reaction was neat and clean, yielding only homogeneous product in satisfactory yield. As the reaction was carried out at pH 6.0, dimerization was kept to a minimum. Furthermore, the peptide–maleimide construct showed high immunogenicity with superior proteolytic stability, a

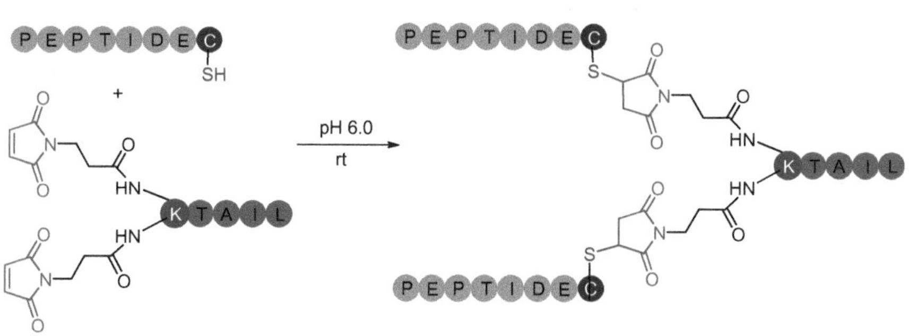

Figure 15 Maleimide linked to a Lys core bearing peptide epitopes to enhance immunogenicity.

property that can be related to its capacity to generate a strong antibody response and also to stimulate T cells. Overall, the method can be considered efficient both on the basis of synthesis and antigenic capacity.

16. TETRA-BRANCHED MALEIMIDO DERIVATIVES AS PLATFORMS FOR VACCINE PREPARATION

Branched peptides, in particular the tetrabranched system, act as ligands for peptidergic receptors. These branched peptides showed high affinity for their receptors, thus drastically reducing the risk of peptidase action (Falciani et al., 2007). Hence, multibranched peptides are of interest as ligands for peptidergic receptors. Guerrini et al. (2014) synthesized tetrabranched derivatives of nociceptin/orphanin FQ in high yield and purity. These derivatives were designed to stimulate food intake. However, we consider that they could also serve as platforms for vaccine development, and hence provide their description below (Fig. 16).

The synthesis of these derivatives involves: (**A**) acetylation of ethylendiamine using Boc-Lys(Boc), deprotection of Boc with TFA and

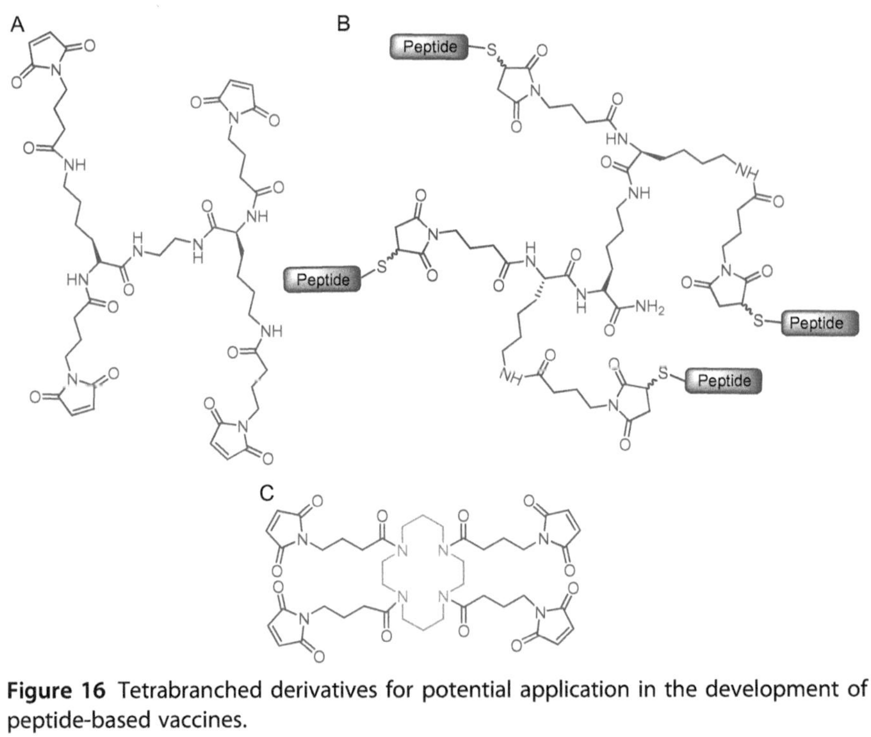

Figure 16 Tetrabranched derivatives for potential application in the development of peptide-based vaccines.

treatment with 4-(2,5-dioxo-2,5-dihydro-1*H*-pyrrol-1-yl)butanoic acid in the presence of *N*-[(dimethylamino)-1*H*-1,2,3-triazolo[4,5-*b*]pyridin-1-ylmethylene]-*N*-methylmethanaminium hexafluorophosphate *N*-oxide (HATU); (**B**) initial reaction of Boc-Lys(Boc) and H-Lys-NH_2 in the presence of 1-(3-dimethylaminopropyl)3-3ethylcarbodiimide (EDCI)/1-hydroxybenzotriazole (HOBt) as coupling reagent; the removal of Boc using TFA and the coupling of the resulting product with maleimido derivative, as above, followed by thiol-Michael addition with a Cys-functionalized peptide using $NaHCO_3$ solution; and (**C**) acylation of cyclam (orange (light gray in the print version) in Fig. 16**C**) as above in **B**. The desired antigenic peptides can also be stapled to all maleimido functions in the remaining systems (**A** and **C**). It should be emphasized here that reaction conditions are mild with respect to solvent, temperature, and pH, and its speed and the purity of the final product attained make this approach more widely applicable to obtain peptidic and non-peptidic compounds.

17. BROMOMALEIMIDE-HEXANOIC ACID AS A LINKER FOR MULTIBRANCHED PEPTIDES

In addition to the abovementioned strategies for the construction of peptide-based vaccines, Ramesh et al. (2015) added a novel platform using 6-(bromomaleimido)hexanoic acid as the bridge. This maleimide moiety enabled the stapling of two peptide epitopes into a bromomaleimide molecule (Fig. 17). This approach contrasts with other maleimide chemistry wherein only one epitope can be tethered per maleimide group (Monso et al., 2013). Also, the new platform was found to be stable for all SPPS conditions. The platform was developed from an economical and efficient route starting from bromomaleic anhydride and amino hexanoic acid. Two distinct peptide epitopes were attached; one at the carboxylic acid end and another at both bromide and nonbromide positions of the maleimide. Using this platform 2^n epitopes were conjugated to *n* number of bromomaleimide molecules, revealing high performance of this moiety. Furthermore, the method also has the advantage of achieving high conversion in less than a minute, together with straightforward purification and characterization. Another merit of this technique is that the final constructs take globular form, thus resembling the shape to be encountered in proteins. Constructs differing in the number of branches, viz two (**A**), four (**B**), and eight (**C**), were assembled with almost negligible dimerization at rt. using a slightly acidic pH.

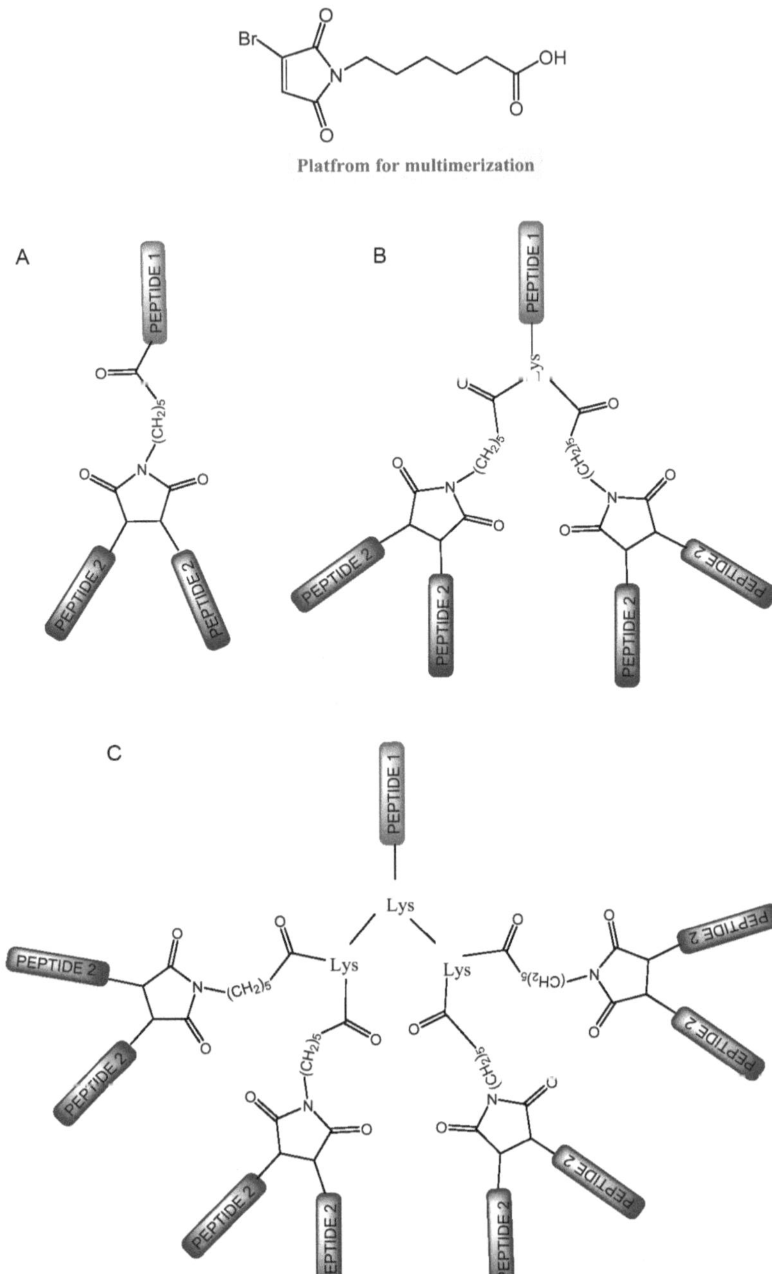

Figure 17 General representation of the multibranched peptides in which two distinct peptide epitopes (blue (dark gray in the print version) and pink (gray in the print version)) are tethered: (**A**) two-armed MAP, (**B**) four-armed MAP, and (**C**) eight-armed MAP.

So far, we have discussed the various platforms to construct peptide vaccines and their advantages and disadvantages. Furthermore, researchers working in vaccine preparation strategies have two options, namely to develop new system(s) or to modify existing ones to achieve a candidate vaccine with high potency. Each of the aforementioned methods could result in an entirely different scenario or may be complementary to one another. Also, distinct methods may not produce the same results, thus they should be evaluated and compared. The foregoing section gives a glimpse of results obtained by various research groups that have evaluated and compared the potency of various chemical platforms for the development of vaccines.

In a study by Monsó et al. (2011) on peptide-based vaccines for CSFV, three peptide-based systems carrying B- and T-cell epitopes were designed by both SPPS and by ligation in solution for comparison purposes (Fig. 18). Various constructs displaying four copies of B-cell epitopes from E2 protein were designed and attached to a T-cell epitope from NS2-3 protein via a polyLys core, along with two extra Lys residues defining a putative cleavage site for cathepsin D, which is believed to be involved in MHC class II antigen processing *in vivo*. In the case of the convergent approach, it was difficult to obtain desirable amounts and purity of the linear peptides to be tethered. In addition, they were insoluble in solvents for ligation. In view of these shortcomings, the authors reverted to SPPS using ChemMatrix® resin as

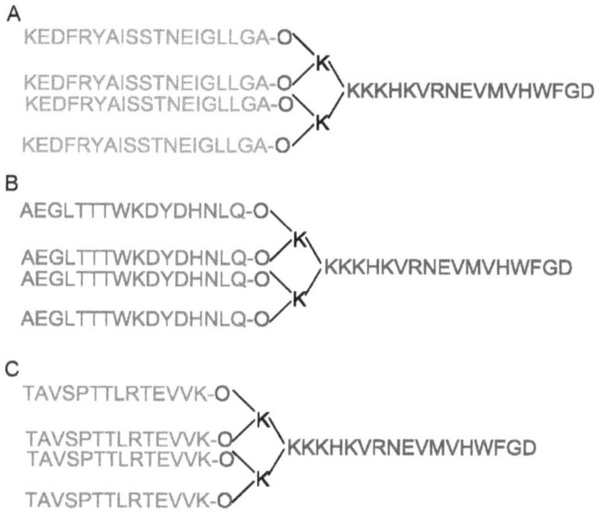

Figure 18 Tetrabranched peptide vaccines for swine fever virus: "O" (green (gray in the print version)) represents 3,6-dioxaoctanoic acid.

the solid support and PEG-like 3,6-dioxaoctanoic acid as spacer to enhance the water solubility of the compound. The synthetic results revealed that stepwise SPPS was more advantageous than ligation in solution. All three dendrimers were tested as vaccine candidates in an experimental pig model. Dendrimer **A** elicited higher antibody titers than the other two dendrimers.

Zeng, Ghosh, Macris, Pagnon, and Jackson (2001) employed T- and B-cell epitopes to construct peptide-based immunogens using a range of chemoselective ligation methods: thioether bond formation, disulfide bond pattern, and oxime bond linkage. The B-cell epitope had the sequence EHWSYGLRPC from luteinising hormone releasing hormone, whereas the T-cell epitope sequence was GALNNRFQIKGVELKS from the light chain (HA2) of influenza virus hemagglutinin. In order to evaluate the efficacy of these vaccines, they were administered to animal models. Also, the effect of stapling the epitope either through its N- or C-terminus was examined. The study revealed that thioether bond formation was the simplest ligation method, thus producing the highest yield. The oxime- and disulfide bond formations required an extra step in the reaction and provided comparatively lower yields. The immunogenicity of the thioether bond- and oxime-containing vaccines exhibited greater activity than disulfide-containing immunogen. Although these vaccine systems were stable at pH 7.0, they demonstrated distinct stabilities *in vivo*, which in turn influence their immunogenicity. The disulfide-containing immunogen showed less activity and was less stable *in vivo*. These observations could be attributed to the fact that this linkage is rarely seen in biological systems. The orientation of the epitopes, i.e., linkage through either the N- or C-terminus, had a small, but not significant, effect on activity. It then became more evident that in spite of using the same epitopes to construct the vaccines, the response varies in function of the chemical approach followed. This observation opens up new avenues to develop tailor-made vaccines to elicit a particular response.

18. COMPARATIVE ACCOUNT OF DIFFERENT PRESENTATION SYSTEMS FOR HIV PEPTIDE IMMUNOGEN

Cruz et al. (2004) used various presentation strategies, such as oligomerization, MAP dendrimers, and conjugation to dextran beads or to other macromolecular carriers, to compare and improve the immunogenicity of an HIV antigenic peptide. They selected RQSTPIGLGQALYTT

(317–331) of the V3 region of HIV-1, JY1 isolate (subtype D) as B-cell epitope, and the QYIKANSKFIGITEL (830–844) region of tetanus toxoid as T-cell epitope. The peptides were flanked by Cys-Gly and Gly-Cys to promote oligomerization. (i) Polymerization of these peptides was carried out via Cys residues protected with Acm by iodine, and polymerization was stopped after 30 min by addition of L-ascorbic acid. After exhaustive dialysis and lyophilization, the polymer was characterized by a HPLC gel filtration technique; (ii) MAP was constructed by stepwise Boc solid-phase strategy in which Cys was protected as 4-methoxybenzyl (Mob). Boc-Lys(Boc) was used for branching; (iii) peptides were assembled on chemically modified dextran beads by converting hydroxyl function of dextran into amino groups. Dextran was used as an adjuvant because it is inexpensive, biocompatible and biodegradable. Next, Fmoc-amino acids were coupled sequentially by DIC/HOBt. Peptide was released from the dextran by treatment with 0.1 M NaOH; (iv) peptides/MAPs were conjugated to hepatitis B surface antigen protein, an effective carrier protein, using phosphate buffer and succinic anhydride. The activated protein was exhaustively dialyzed, filtered, and characterized. The results of immunization revealed that MAP-carrier protein conjugates had higher antipeptide titers than other presentation strategies. Compared to monomeric peptide, the dendrimer version showed enhancement in cross-reactivity toward heterologous V3 sequences. Results also showed that a small amount of antigen gave a largely responsive antibody. Thus this comparative account would find broad application in vaccine preparation.

19. MERITS AND DEMERITS OF CHEMOSELECTIVE AND SPPS REACTIONS: M2e EPITOPE OF INFLUENZA VIRUS AS A CASE STUDY

Two of the most useful techniques for vaccine construction are chemoselective thioether ligation and SPPS. In view of the few reports on comparative study of solution and stepwise SPPS, Kowalczyk, de la Torre, and Andreu (2010) undertook to chemically describe the pros and cons of these methods, using a difficult model, namely the M2e epitope of influenza virus. Three MAP models were designed and constructed (Fig. 19) either by ligation or SPPS. All the peptides were assembled on a synthesizer using the Fmoc/*t*Bu protocol, and branching was achieved by inserting Fmoc-Lys(Fmoc) at appropriate points. For chemoselective ligation, a standard Cl-Ac Lys core and thiol peptide were used. M2e peptide

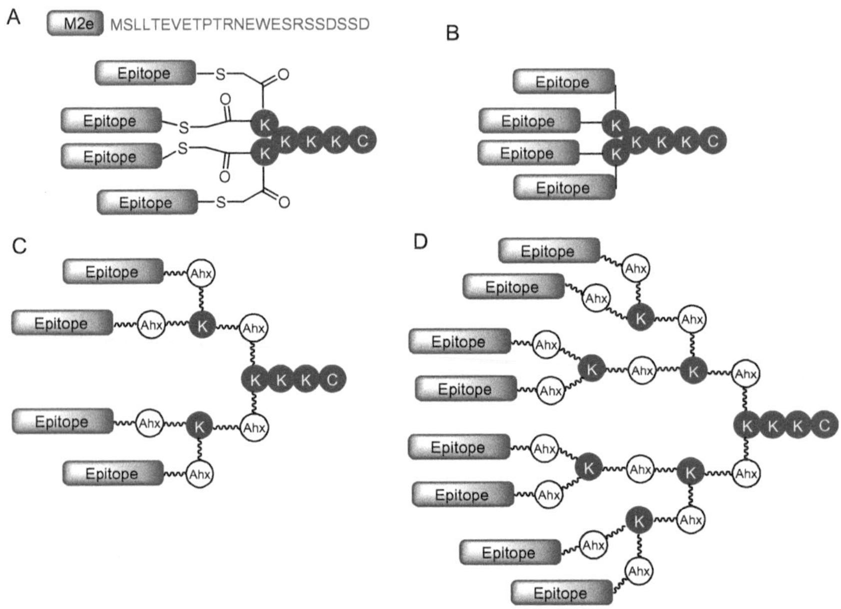

Figure 19 General structures of the dendrimers synthesized in this study (Ahx, aminohexanoic acid): (**A**) thioether bond between the Cl-Ac Lys core and thiol peptide, (**B**) line indicates amide bond, and (**C** and **D**) wavy lines around Ahx indicate the flexibility provided by this residue

was problematic during the synthesis, leading to unsatisfactory results. Changing the consecutive Ser residues in the sequence to pseudoproline dipeptide gave an acceptable product. However, the HPLC trace of the MAPs was of poor resolution, had considerable levels of dimer, and even tetramer of the epitopes, as revealed by mass spectrometry. In this event, the authors switched to SPPS by tailoring the M2e epitope and also enhanced the flexibility of the dendrimer by introducing aminohexanoic acid as spacer unit at each branching point. This strategy/model was readily accessible and trouble-free not only for tetravalent but also for octavalent MAPs. This study revealed that careful fine-tuning of synthetic conditions is essential to achieve chemically well-defined MAPs.

20. SUMMARY AND OUTLOOK

Humans are affected by numerous devastating diseases. To address this critical issue, traditional vaccines use attenuated or a live forms of pathogens. Although this method has been successful, several difficulties have often been encountered to maintain the efficacy of these vaccines and to ensure their safety on a lot-by-lot basis. On the other hand, the use of synthetic versions of immunogens have heralded a new era in the preparation of peptide-based vaccines that specifically induce a boost in immune system response. Since its inception, this field has experienced several advancements, such as the development of more effective and less toxic adjuvants, the selection of epitopes that resemble the 3D structure of the original proteins, the design and development of new scaffolds for different presentation strategies, and also the development of new routes of administration/immunization. The peptide approach is not free of practical and theoretical difficulties, the major one being the conformational B-cell epitopes, which are difficult to mimic by a simple synthetic analog.

A key hit in the peptide-vaccine field was the development of MAPs by Tam in 1988, which led to the invention of several other presentation strategies that make up the arsenal of vaccine preparation tools. These bibliographies reveal the intricacies in the development of such vaccines and the number of parameters—approach, multivalency, spacers, linkers, chemical ligation methods, and selection and combination of B- and T-cell epitopes, among others—that influence the conformation and immunogenicity of peptides for vaccine purposes. Although great progress has been made in the peptide-based vaccine field, chemists, immunologists, and clinicians continue to work together to fuel the momentum, and thus bring about solutions for public health.

ACKNOWLEDGMENT

The work in the laboratory of authors was partially funded in part by the following: the National Research Foundation, the University of KwaZulu-Natal, Asphen Pharmacare (South Africa), SENESCYT (Ecuador), the CICYT (CTQ2012-30930), the *Generalitat de Catalunya* (2014 SGR 137), and the Institute for Research in Biomedicine Barcelona (IRB Barcelona) (Spain).

REFERENCES

Banchereau, J., & Steinman, R. M. (1998). Dendritic cells and the control of immunity. *Nature*, *392*, 245–252.

Boas, U., & Heegaard, P. M. H. (2004). Dendrimers in drug research. *Chemical Society Reviews*, *33*, 43–63.

Brumeanu, T.-D., Casares, S., Bot, A., Bot, S., & Bona, C. A. (1997). Immunogenicity of a contiguous T-B synthetic epitope of the A/PR/8/34 influenza virus. *Journal of Virology*, *71*, 5473–5480.

Chang, C.-Y., Huang, C.-C., Deng, M.-C., Huang, Y.-L., Lin, Y.-J., Liu, H.-M., et al. (2012). Antigenic mimicking with cysteine-based cyclized peptides reveals a previously unknown antigenic determinant on E2 glycoprotein of classical swine fever virus. *Virus Research*, *163*, 190–196.

Chen, Y.-S., Hung, Y.-C., Lin, W.-H., & Huang, G. S. (2010). Assessment of gold nanoparticles as a size-dependent vaccine carrier for enhancing the antibody response against synthetic foot-and-mouth disease virus peptide. *Nanotechnology*, *21*, 195101/195101–195101/195108.

Cruz, L. J., Iglesias, E., Aguilar, J. C., Cabrales, A., Reyes, O., & Andreu, D. (2004). Different immune response of mice immunized with conjugates containing multiple copies of either consensus or mixotope versions of the V3 loop peptide from human immunodeficiency virus type 1. *Bioconjugate Chemistry*, *15*, 1110–1117.

Drijfhout, J. W., & Bloemhoff, W. (1991). A new synthetic functionalized antigen carrier. *International Journal of Peptide and Protein Research*, *37*, 27–32.

Esser, M. T., Marchese, R. D., Kierstead, L. S., Tussey, L. G., Wang, F., Chirmule, N., et al. (2003). Memory T cells and vaccines. *Vaccine*, *21*, 419–430.

Falciani, C., Lozzi, L., Pini, A., Corti, F., Fabbrini, M., Bernini, A., et al. (2007). Molecular basis of branched peptides resistance to enzyme proteolysis. *Chemical Biology and Drug Design*, *69*, 216–221.

Frisch, B., Boeckler, C., & Schuber, F. (1996). Synthesis of short polyoxyethylene-based heterobifunctional cross-linking reagents. Application to the coupling of peptides to liposomes. *Bioconjugate Chemistry*, 7, 180–186.

Gomara, M. J., Fernandez, L., Perez, T., Ercilla, G., & Haro, I. (2010). Assessment of synthetic chimeric multiple antigenic peptides for diagnosis of GB virus C infection. *Analytical Biochemistry*, *396*, 51–58.

Grandjean, C., Rommens, C., Gras-Masse, H., & Melnyk, O. (2000). One-pot synthesis of antigen-bearing, lysine-based cluster mannosides using two orthogonal chemoselective ligation reactions. *Angewandte Chemie, International Edition*, *39*, 1068–1072.

Guerrini, R., Marzola, E., Trapella, C., Pela, M., Molinari, S., Cerlesi, M. C., et al. (2014). A novel and facile synthesis of tetra branched derivatives of nociceptin/orphanin FQ. *Bioorganic and Medicinal Chemistry*, *22*, 3703–3712.

Hahn, K. W., Klis, W. A., & Stewart, J. M. (1990). Design and synthesis of a peptide having chymotrypsin-like esterase activity. *Science*, *248*, 1544–1547.

Haro, I., Perez, S., Garcia, M., Chan, W. C., & Ercilla, G. (2003). Liposome entrapment and immunogenic studies of a synthetic lipophilic multiple antigenic peptide bearing VP1 and VP3 domains of the hepatitis A virus: A robust method for vaccine design. *FEBS Letters*, *540*, 133–140.

Hilleman, M. R. (2000). Vaccines in historic evolution and perspective: A narrative of vaccine discoveries. *Vaccine*, *18*, 1436–1447.

Houseman, B. T., Gawalt, E. S., & Mrksich, M. (2003). Maleimide-functionalized self-assembled monolayers for the preparation of peptide and carbohydrate biochips. *Langmuir*, *19*, 1522–1531.

Jackson, D. C., Lau, Y. F., Le, T., Suhrbier, A., Deliyannis, G., Cheers, C., et al. (2004). A totally synthetic vaccine of generic structure that targets toll-like receptor 2 on dendritic cells and promotes antibody or cytotoxic T cell responses. *Proceedings of the National Academy of Sciences of the United States of America*, *101*, 15440–15445.

Kent, S. B. H. (1988). Chemical synthesis of peptides and proteins. *Annual Review of Biochemistry*, *57*, 957–990.

Kowalczyk, W., de la Torre, B. G., & Andreu, D. (2010). Strategies and limitations in dendrimeric immunogen synthesis: The influenza virus M2e epitope as a case study. *Bioconjugate Chemistry*, *21*, 102–110.

Langhans, B., Schweitzer, S., Nischalke, H. D., Braunschweiger, I., Sauerbruch, T., & Spengler, U. (2004). Hepatitis C virus-derived lipopeptides differentially induce epitope-specific immune responses in vitro. *Journal of Infectious Diseases*, *189*, 248–253.

Lerner, R. A. (1982). Tapping the immunological repertoire to produce antibodies of predetermined specificity. *Nature, 299*, 592–596.

Li, W., Joshi, M., Singhania, S., Ramsey, K., & Murthy, A. (2014). Peptide vaccine: Progress and challenges. *Vaccines, 2*, 515–536.

Liu, T.-Y., Hussein, W. M., Jia, Z., Ziora, Z. M., McMillan, N. A. J., Monteiro, M. J., et al. (2013). Self-adjuvanting polymer-peptide conjugates as therapeutic vaccine candidates against cervical cancer. *Biomacromolecules, 14*, 2798–2806.

Liu, T.-Y., Hussein, W. M., Toth, I., & Skwarczynski, M. (2012). Advances in peptide-based human papillomavirus therapeutic vaccines. *Current Topics in Medicinal Chemistry, 12*, 1581–1592.

Martin, F. J., & Papahadjopoulos, D. (1982). Irreversible coupling of immunoglobulin fragments to preformed vesicles: An improved method for liposome targeting. *Journal of Biological Chemistry, 257*, 286–288.

Medzhitov, R., & Janeway, C., Jr. (2000). Innate immune recognition: Mechanisms and pathways. *Immunology Reviews, 173*, 89–97.

Monso, M., de la Torre, B. G., Blanco, E., Moreno, N., & Andreu, D. (2013). Influence of conjugation chemistry and B epitope orientation on the immune response of branched peptide antigens. *Bioconjugate Chemistry, 24*, 578–585.

Monso, M., Kowalczyk, W., Andreu, D., & de la Torre, B. G. (2012). Reverse thioether ligation route to multimeric peptide antigens. *Organic and Biomolecular Chemistry, 10*, 3116–3121.

Monsó, M., Tarradas, J., de la Torre, B. G., Sobrino, F., Ganges, L., & Andreu, D. (2011). Peptide vaccine candidates against classical swine fever virus: T cell and neutralizing antibody responses of dendrimers displaying E2 and NS2–3 epitopes. *Journal of Peptide Science, 17*, 24–31.

Montal, M., Montal, M. S., & Tomich, J. M. (1990). Synporins - synthetic proteins that emulate the pore structure of biological ionic channels. *Proceedings of the National Academy of Sciences of the United States of America, 87*, 6929–6933.

Noren, C. J., Anthony-Cahill, S. J., Griffith, M. C., & Schultz, P. G. (1989). A general method for site-specific incorporation of unnatural amino acids into proteins. *Science, 244*, 182–188.

Patarroyo, M. E., & Patarroyo, M. A. (2008). Emerging rules for subunit-based, multiantigenic, multistage chemically synthesized vaccines. *Accounts of Chemical Research, 41*, 377–386.

Purcell, A. W., McCluskey, J., & Rossjohn, J. (2007). More than one reason to rethink the use of peptides in vaccine design. *Nature Reviews Drug Discovery, 6*, 404–414.

Ramesh, S., Cherkupally, P., Govender, T., Kruger, H. G., Albericio, F., & de la Torre, B. G. (2015). 6-(Bromomaleimido)hexanoic acid as a connector for the construction of multiple branched peptide platforms. *Organic Letters, 17*, 464–467.

Rao, C., & Tam, J. P. (1994). Synthesis of peptide dendrimer. *Journal of the American Chemical Society, 116*, 6975–6976.

Roberts, M. J., Bentley, M. D., & Harris, J. M. (2002). Chemistry for peptide and protein PEGylation. *Advanced Drug Delivery Reviews, 54*, 459–476.

Rose, K. (1994). Facile synthesis of homogeneous artificial proteins. *Journal of the American Chemical Society, 116*, 30–33.

Rose, K., Zeng, W., Brown, L. E., & Jackson, D. C. (1995). A synthetic peptide-based polyoxime vaccine construct of high purity and activity. *Molecular Immunology, 32*, 1031–1037.

Rueckert, C., & Guzman, C. A. (2012). Vaccines: From empirical development to rational design. *PLoS Pathogens, 8*, e1003001.

Samayoa, L., Diaz-Mitoma, F., & Azizi, A. (2011). Characterization of a branched lipopeptide candidate vaccine against influenza A/Puerto rico 8/34 which is recognized by human B and T-cell immune responses. *Virology Journal, 8*, 309.

Schnoelzer, M., & Kent, S. B. H. (1992). Constructing proteins by dovetailing unprotected synthetic peptides: Backbone-engineered HIV protease. *Science*, *256*, 221–225.

Skwarczynski, M., & Toth, I. (2011). Peptide-based subunit nanovaccines. *Current Drug Delivery*, *8*, 282–289.

Skwarczynski, M., Zaman, M., Urbani, C. N., Lin, I. C., Jia, Z., Batzloff, M. R., et al. (2010). Polyacrylate dendrimer nanoparticles: A self-adjuvanting vaccine delivery system. *Angewandte Chemie, International Edition*, *49*, 5742–5745.

Smith, M. (1985). In vitro mutagenesis. *Annual Review of Genetics*, *19*, 423–462.

Steinman, R. M., Pack, M., & Inaba, K. (1997). Dendritic cells in the T-cell areas of lymphoid organs. *Immunology Reviews*, *156*, 25–37.

Tam, J. P. (1988). Synthetic peptide vaccine design: Synthesis and properties of a high-density multiple antigenic peptide system. *Proceedings of the National Academy of Sciences of the United States of America*, *85*, 5409–5413.

Tam, J. P., & Zavala, F. (1989). Multiple antigen peptide: A novel approach to increase detection sensitivity of synthetic peptides in solid-phase immunoassays. *Journal of Immunological Methods*, *124*, 53–61.

Timmerman, P., Puijk, W. C., & Meloen, R. H. (2007). Functional reconstruction and synthetic mimicry of a conformational epitope using CLIPS technology. *Journal of Molecular Recognition*, *20*, 283–299.

Weiland, E., Stark, R., Haas, B., Ruemenapf, T., Meyers, G., & Thiel, H. J. (1990). Pestivirus glycoprotein which induces neutralizing antibodies forms part of a disulfide-linked heterodimer. *Journal of Virology*, *64*, 3563–3569.

Wells, J. A., Vasser, M., & Powers, D. B. (1985). Cassette mutagenesis: An efficient method for generation of multiple mutations at defined sites. *Gene*, *34*, 315–323.

Wlodawer, A., Miller, M., Jaskolski, M., Sathyanarayana, B. K., Baldwin, E., Weber, I. T., et al. (1989). Conserved folding in retroviral proteases: Crystal structure of a synthetic HIV-1 protease. *Science*, *245*, 616–621.

Zaman, M., Skwarczynski, M., Malcolm, J. M., Urbani, C. N., Jia, Z., Batzloff, M. R., et al. (2011). Self-adjuvanting polyacrylic nanoparticulate delivery system for group A streptococcus (GAS) vaccine. *Nanomedicine Nanotechnology*, *7*, 168–173.

Zeng, W., Ghosh, S., Macris, M., Pagnon, J., & Jackson, D. C. (2001). Assembly of synthetic peptide vaccines by chemoselective ligation of epitopes: Influence of different chemical linkages and epitope orientations on biological activity. *Vaccine*, *19*, 3843–3852.

CHAPTER SIX

Understanding Amino Acid Mutations in Hepatitis B Virus Proteins for Rational Design of Vaccines and Drugs

Ke Shen*,[1], Li Shen*,†,[1], Jing Wang*, Zhi Jiang*,‡,[2], Bairong Shen*,[2]
*Center for Systems Biology, School of Computer Science and Technology, Soochow University, Suzhou, 215006 PR China
†Institute of Biological Sciences and Biotechnology, Donghua University, Shanghai, 201620 PR China
‡School of Medicine, Soochow University, Suzhou, 215123 PR China
[2]Corresponding authors: e-mail address: szjiangzhi@163.com; bairong.shen@suda.edu.cn

Contents

Abstract

The hepatitis B virus (HBV) genome encodes four proteins, i.e., DNA polymerase, surface protein, X, and core proteins. HBV undergoes different selective pressures for drug resistance and immune/vaccine escape and mutations are common for the HBV proteins. We here collected all the reported amino acid mutations happened in these four HBV proteins and studied their patterns. The relationship between the mutations and epitopic functions are investigated with bioinformatics tools, based on their sequence information. Some interesting results are observed for the mutation patterns, such as we found the serine and threonine are both for frequently mutated residues and mutant residues, while the tryptophan and methionine have low mutability. The results provide important information for the understanding of the molecular mechanism of virus evolution and therefore will facilitate the future rational design of HBV vaccines or drugs.

[1] Contribute equally to the work.

Advances in Protein Chemistry and Structural Biology, Volume 99
ISSN 1876-1623
http://dx.doi.org/10.1016/bs.apcsb.2015.03.006

1. INTRODUCTION

Hepatitis B virus (HBV) is a DNA virus and has chronically infected more than 350 million people. The chronic infection may lead to hepatic decompensation, cirrhosis (Jeng, Sheen, & Liaw, 2010; Kim et al., 2009), hepatocellular carcinoma, liver failure, and even death (Chen et al., 2006; Liaw & Chu, 2009; Shepard, Simard, Finelli, Flore, & Bell, 2006; Zhou, Luo, Chen, & Gong, 2015). Therefore, it is very necessary to develop effective drugs or vaccines to improve the global health. We checked web of science (core collection) for the studies on HBV and their mutations in recent 10 years. The research profiles for HBV and HBV mutations are shown in Fig. 1. Lots of researches about HBV (Fig. 1A) mainly focused on the virus infection and replication. In epidemiology, most of the reported works are about the prevalence, the risk of disease like hepatitis B and hepatocellular carcinoma. Gene mutations and expression are the key topics for the understanding of molecular mechanisms of virus–host interactions. For the antiviral treatment of HBV infections, lamivudine and surface antigen are intensively investigated. As for the HBV mutations (Fig. 1B), most of the research topics are quite similar as the HBV research profile, but more works are about the basal core promoter or precore mutations and the fulminant hepatitis. Although there are many amino acid mutations reported associated with the four HBV proteins, i.e., DNA polymerase, surface protein, X, and core proteins. Few work until now has been reported to summarize all the amino acid mutations distributed in these HBV proteins. It is well known that the HBV is constantly mutated evolutionarily to develop resistance to the antivirus drug or escape the host immune responses. Therefore to study the pattern of mutations is necessary to understand the molecular mechanism of virus–host interaction, to know the functions of these proteins in the virus infection and replication, then to develop rationally vaccines and drugs to prevent the transmission of HBV, and therefore the progression of the disease.

The amino acid mutations happened in the HBV proteins are mainly caused by the evolutionary pressures of host immune responses, the effects of antivirus vaccines and drugs. The interactions between virus and the antiviral drug/vaccine are very complex, and heterogeneous. The virus, the environments, and the host responses are diverse for different regions, different periods, or different patients. Systems biological investigation of the complex network between the host genes, virus, and the drugs/vaccines is necessary (Chen, Wang, Guo, & Shen, 2012).

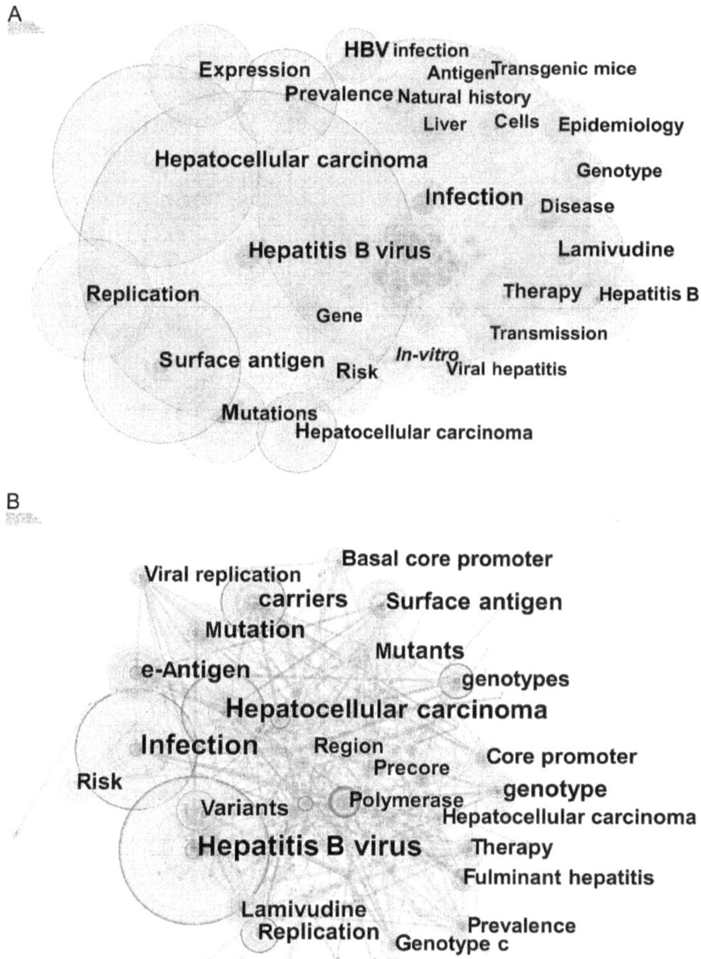

Figure 1 Scientific mapping for the analysis of publications on HBV (A) and HBV mutations (B) with CiteSpace (Chen, 2006).

Global analyzing or summarizing all the amino acid mutations in HBV proteins will facilitate the prediction of the evolution of the epitopic hot regions of the antigens and be helpful to the understanding of complex network between the viral genes, host immune systems, and the antiviral vaccine/drugs, and therefore for vaccine strain recommendations (Du et al., 2012).

In order to investigate the drug resistance and the vaccine escape associated amino acid mutations, we can apply experimental methods to sequence the expression of genes (Huang et al., 2012), to measure the effect of mutants on the binding of drugs to proteins, or we could apply model animals to investigate the changes of the immune responses affected by new mutants (Kamili et al., 2009); however, the experimental methods are costing and time consuming, and some of the complex phenomenon even can only be investigated with computational simulations. With bioinformatics or computational simulation, it is possible to study the effect of mutations based on the available biological data, such as the protein sequence and structure information. Until now, the evolutionary pathway and the changing path of the amino acid mutations in HBV proteins is still unclear and need to be investigated, but this is very difficult since we need to collect the longitudinal and paired samples (Osiowy, Giles, Tanaka, Mizokami, & Minuk, 2006). In this work, we aimed to investigate the global pattern of the amino acid mutations in the HBV proteins and provide a global view of the drug resistance or immune escape strategies of HBV.

At present, the HBV genome has been sequenced and the reference sequences of all the proteins are known. A few databases have been built for the storing and the annotation of the HBV related data, such as, the HBV reverse transcriptase sequence variant database (HBVrtDB, http://hivdb.stanford.edu/HBV/releaseNotes/), which was developed by Rhee et al., stored the mutations happened in reverse transcriptase (Rhee et al., 2010). The other database, HBV knowledge database (HBVdb, https://hbvdb.ibcp.fr/HBVdb/), provides both data and tools for annotation, genotyping, and drug resistance profiling (Hayer et al., 2013).

In the other aspect, many bioinformatics tools have also been developed to understand amino acid mutations happened in virus proteins (Chen & Shen, 2009; Chen, Wang, Guo, & Shen, 2012), such as SHIF, SNP3d, ConSurf, etc. are well-known tools for the analysis of effect of mutations on protein conservation, many bioinformatics tools are developed to predict the MHC class I/II binder (Altuvia & Margalit, 2004; Bhasin & Raghava, 2003; Bordner & Abagyan, 2006; Cui et al., 2006; Donnes & Kohlbacher, 2006; Doytchinova & Flower, 2006; Goodswen, Kennedy, & Ellis, 2014; Guan, Doytchinova, Zygouri, & Flower, 2003; Hammer, 1995; Molero-Abraham, Lafuente, & Reche, 2014; Shen et al., 2014; Wan et al., 2006) and B cell epitope (El-Manzalawy & Honavar, 2010; Greenbaum et al., 2007; Kringelum, Lundegaard, Lund, & Nielsen, 2012; Lian, Ge, & Pan, 2014; Singh, Ansari, & Raghava, 2013; Sweredoski & Baldi, 2008; Yao, Zheng, Liang, & Zhang, 2013; Zhao, Wong, Lu, Hoi, & Li, 2012), with the protein sequence or structure as the input data. There are also immune epitope database established and provided resource for the study of adaptive immune responses (Vita et al., 2015; Zhang et al., 2008).

All of these provide us the chance to annotate the amino acid mutations in the HBV proteins, to identify the evolutionary tendency of the mutations, and then to provide important hints for the rational design of vaccines or drugs.

2. MUTATION DATA AND DISTRIBUTION ANALYSIS

2.1 Data Collection

To collect the amino acid mutations happened in the HBV proteins, we searched both the PUBMED and HBV knowledge databases. The mutations for the four HBV proteins are summarized and presented in Figs. 2–4. Totally there are 526, 372, 131, and 87 amino acid mutations for HBV DNA polymerase, S, X, and core proteins, respectively. The distribution of the mutations in different protein domains and the corresponding domain information are listed in Table 1.

2.2 Distribution of Amino Acid Mutations

It is clear that the amino acid mutations are distributed differently in the different domains of the four HBV proteins. The order of the number is correlated to the selective pressure these proteins underwent. Most of the mutations are happened in HBV DNA polymerase and the S protein since the DNA polymerase undergo pressures from the antiretroviral drugs such as Lamivudine, one of the reverse-transcriptase inhibitors, while the S protein, which is the main surface antigen of HBV and locates in the frontier of the virus–host interaction. S protein is essential to the virus entry to the host, and the selective pressure comes from the escape of the immune attack by the host immune system or the antiviral vaccines. The X and Core proteins have relatively less mutations, since their selective pressure is not strong as the former two proteins. The distribution of amino acid mutations are displayed in Figs. 5–8. The bold lines and the fine lines in Figs. 5–7 represent the distribution of mutated amino acids and mutants, respectively.

2.2.1 Distribution of Mutated Amino Acids

Figure 5 shows the relative mutability of amino acids in the four HBV proteins based on the 20-letter alphabet amino acids. It is clear that the mutation patterns are different for the four HBV proteins since their functions are diverse, but some consensus tendency could be identified from the four mutation patterns, the serine (for DNA polymerase, X, and core proteins) and threonine (for S protein and DNA polymerase) are the most mutated

Hepatitis B virus DNA polymerase (843)

NCBI Reference Sequence: NP_647604.2

>gi|57021117|ref|NP_647604.2| Polymerase [Hepatitis B virus]

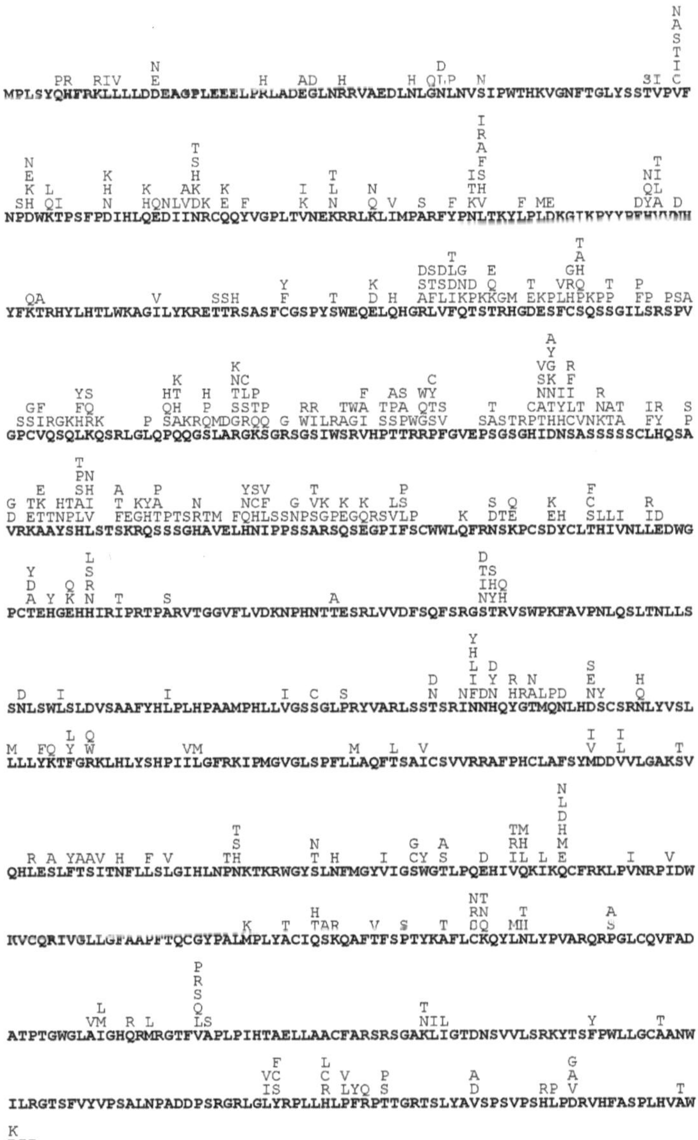

Figure 2 Amino acid mutations distributed in hepatitis B virus DNA polymerase.

Hepatitis B virus S protein (400)

NCBI Reference Sequence: YP_355333.1

```
>gi|77680739|ref|YP_355333.1| large S protein [Hepatitis B virus]

                                      D
                                      S              M
   R                                  R        K  P  Q
   P P T RE                           A       TI  Q  A       G
  ALLWTVPLQ WKQ H TS     L G     LKR  TES   HLTKH TN TESHQ  AGSY L LI
MGGWSSKPRKGMGTNLSVPNPLGFFPDHQLDPAFGANSNNPDWDFNPNKDHWPDANKVGVGAFGPGFTPP

                    T                                  S
             V      V                                  N  L
             T A    D         L                  T     I  H
  NV         MMK   VI         V      V           V    AQ  R
  SI      S  LIQSLSTN     A  RT  K   LT    NS    IK   SA  NT LN K KA
HGGLLGWSPQAQGILTTVPAAPPPASTNRQSGRQPTPISPPLRDTHPQAMQWNSTTFHQALQDPRVRGLY

    K              S
    E     A        L            M
 S  A    AL        AL   SLT     T     S        M
LL  S   EIQS AQNIV HTL TLAKI GLVLT DNIA  L  T  G   VCSS  K    LK
FPAGGSSSGTVNPVPTTASPISSIFSRTGDPAPNMESTTSGFLGPLLVLQAGFFLLTRILTIPQSLDSWW

        L                 S
        V K               A
        P A H             V    N                               F
        S G R             M    A                               S
   S   ETTV L L LL QI S LLIC   IR   H  Y   S   IC        SVGMVRCR   L
TSLNFLGGAPTCPGQNSQSPTSNHSPTSCPPTCPGYRWMCLRRFIIFLFILLLCLIFLLVLLDYQGMLPV

                           H
                  V    N   I      R
           M S  A ISS NRK  L      Q  LHA            E
 P         ARG PNRAAL REI LN I   IWLPAK W      P    V S  D
GQ I RSTAN VETSKISMTTVGDNFTY Y  RSESTER S  T  AV   LGKY  PLVLAH   F S
CPLLPGTSTTSTGPCRTCTIPAQGTSMFPSCCCTKPSDGNCTCIPIPSSWAFARFLWEWASVRFSWLSLL

            A      F       K  H  K  S
            V  A T S TS    QR  R  R  M      L    ■
AQLG  CARSL I  FLVML ILF  RN CSTVN  IQ  A SCY  AS
VPFVQWFVGLSPTVWLSAIWMMWYWGPSLYNILSPFLPLLPIFFCLWVYI

■ indicates deletion
```

Figure 3 Amino acid mutations distributed in hepatitis B virus S protein.

residues, the next frequently mutated residues are alanine (for X protein), proline (for S, X, and core proteins), for all the four proteins, the tryptophan, and methionine residues are very less mutated. For functional analysis of the mutations, we classified the amino acids to six groups, i.e., hydrophobic (V I L F M W Y C), polar (N Q S), positively charged (R K H), negatively charged (D E), conformational (G P), and (A T) groups (Shen & Vihinen, 2004), based on their physicochemical similarity and properties. Figure 6 presents the mutations according to the six amino acid groups. The tendency is clear for all the HBV proteins that hydrophobic groups are the hot spots for mutations, but the negatively charged residues are less mutated. Figure 7 summarized the overall tendency of the mutations at the amino acid or physicochemical group level, the most frequently mutated residues are serine, threonine, and proline, the less mutated residue are tryptophan and methionine. In the other aspect, the hydrophobic, polar, and (A T) groups have the higher relative mutability, while the others, especially the negative charged group, has low variability.

Hepatitis B virus X protein（154）

NCBI Reference Sequence: NP_647606.1

>gi|21326587|ref|NP_647606.1| X protein [Hepatitis B virus]

```
                                   G       T
                                   S     T L
                             I     D     L G
    M      T             S   FA    PI  S A I  SVN
    VYT  RSS        LS   C QSLPRSF AVSPAPPPV STND S         TS   T  T
MAARLCCQLDPARDVLCLRPVGAESRGRPVSGPLGTLPSPSSSAVPADHGAHLSLRGLPVCAFSSAGPCA

                LH                                      F
                MA                                      N
                WS                                      S
               SHF            FT               A        L
       Y       PGI  R  R      AG  I G  K       N        V  QL
     F G A AM TRRN  TQ Y   I  PVRIM WI ETIS  L TE   S NDTM MIY    T Q
LRFTSARRMETTVNAHQVLPKVLHKRTLGLSAMSTTDLEAYFKDCVFKDWEELGEEIRLKVFVLGGCRHK

   V
  RA SS
LVCSPAPCNFFTSA
```

Hepatitis B virus precore/core protein（212）

NCBI Reference Sequence: YP_355335.1

>gi|77680741|ref|YP_355335.1| precore/core protein [Hepatitis B virus]

```
                                                 T             T
L         F T S F           DL    T      TA       A  Y  V       A  F DS
MQLFHLCLIISCSCPTVQASKLCLGWLWGMDIDPYKEFGASVELLSFLPSDFFPSIRDLLDTASALYREA

                                T
                                G     V
                         S      V  D  T      G         L
   SD  TAN   I   VV  VDVIT SS   N  Q QIA DAI N   THV I F  I
LESPEHCSPHHTALRQAILCWGELMNLATWVGSNLEDPASRELVVSYVNVNMGLKIRQLLWFHISCLTFG

                  L
 Q  V             Q    S           A   C                         A  G
KDV I             TP   Q          NC I Q C T                     K PAP
RETVLEYLVSFGVWIRTPPAYRPPNAPILSTLPETTVVRRRGRSPRRRTPSPRRRRSQSPRRRRSQSRES

QC
```

Figure 4 Amino acid mutations distributed in hepatitis B virus X and core proteins.

2.2.2 Spectrum of the HBV Protein Mutants

To study the spectrum of HBV mutants, it can be seen from the Figs. 5–7 that the mutation patterns are different for the four HBV proteins, but some trends can still be identified which is similar to the patterns in mutated amino acid profiles, such as serine (for DNA polymerase, S, and X proteins) and threonine (for all the four proteins) are the top mutant residues, alanine is the next one for the frequently mutant residue. In addition, tryptophan and methionine not only have the lowest mutability but are also less frequently mutant residues. The observations are interesting and they indicate that these amino acids are important to the functions of HBV proteins, if these residues mutated, some other residues need to be mutated back to them to keep the functionality of the whole proteins. The molecular

Table 1 Distribution of Amino Acid Mutations in Different Domains of HBV Proteins

HBV Protein	Domain	Location	Pfam ID (Sequence Identity)	Number of Mutations
DNA polymerase	DNA_pol_viral_N	1→352	00242 (78%)	357
	RVT_1	381→600	00078 (64%)	76
	RT_like	516→600	cl02808	25
	DNA_pol_viral_C	601→843	00336 (90%)	81
S protein	vMSA	30→395	00695 (74%)	347
X protein	X	1→142	00739 (87%)	125
Core protein	Hep_core_N	1→27	08290 (96%)	6
	Hepatitis_core	31→178	00906 (89%)	69

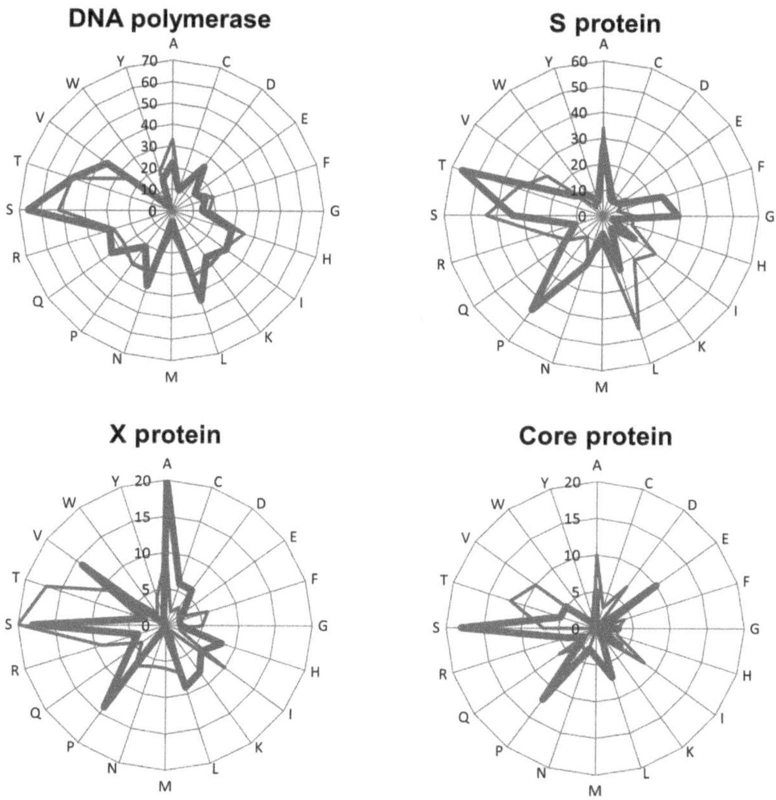

Figure 5 Relative mutability of mutated (bold line) and mutant residues (fine line) for the four HBV proteins based on 20-letter alphabet amino acids.

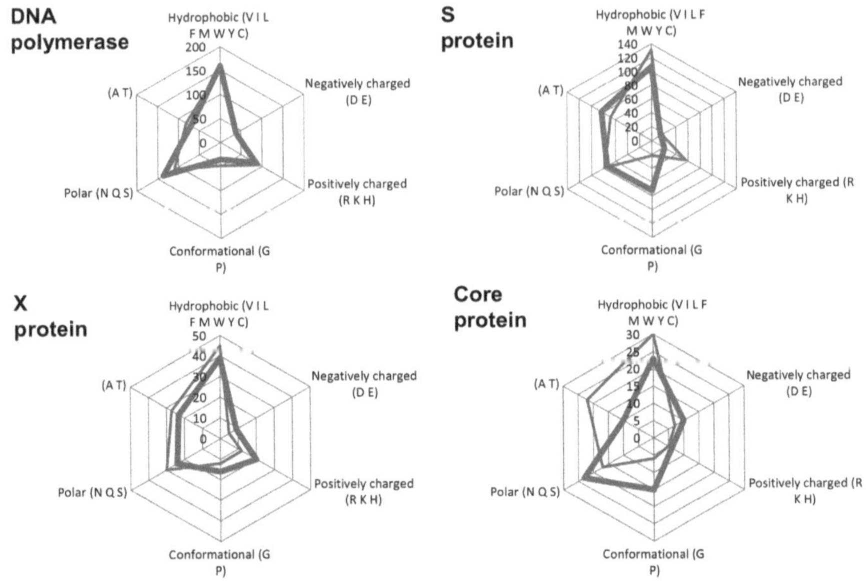

Figure 6 Relative mutability of mutated (bold line) and mutant residues (fine line) for the four HBV proteins based on 6-group alphabet amino acids.

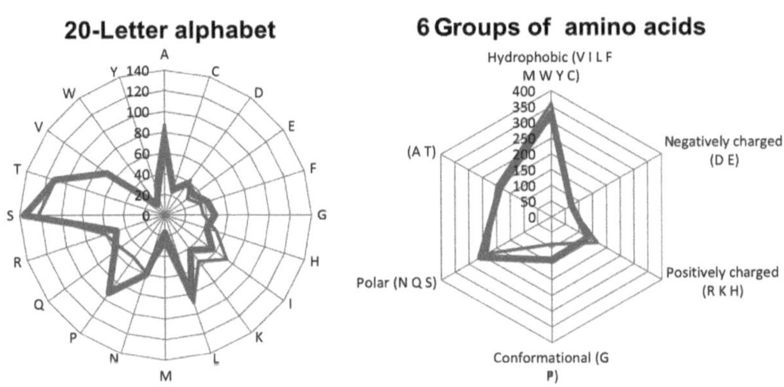

Figure 7 The relative mutability of mutated (bold line) and mutant residues (fine line) for the all the HBV proteins based on 20-letter (left) and 6-group (right) alphabet amino acids.

mechanism needs to be further investigated to understand the observations. According to the patterns displayed in Fig. 6, the profiles of the physicochemical properties of the mutated groups and the mutant groups are similar for HBV DNA polymerase and X proteins, for the former protein, the hydrophobic, polar, and positively charged amino acid groups have higher

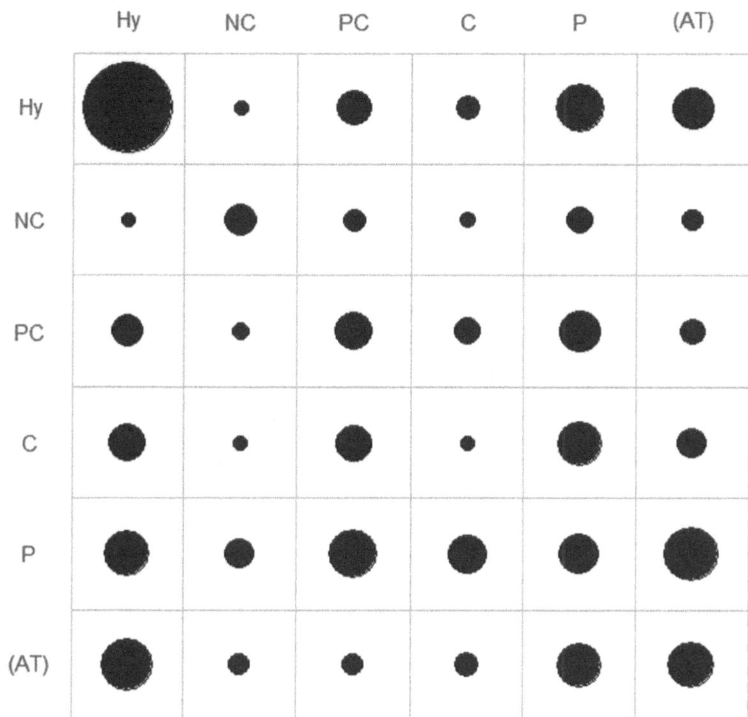

Figure 8 The mutation matrix circles. Rows are the mutant amino acid groups and the columns are mutated amino acid groups. The sizes of the circles correlate with the number of mutations.

mutability and the other three groups have lower mutability, both for mutated residues and mutant residues. For the late protein, the hydrophobic, (A T), and polar amino acid groups have higher mutability and the other three groups have lower mutability, both for mutated residues and mutant residues. The S and core proteins have different profiles for the mutated and mutant residues.

As a summary, the patterns for the amino acid mutations are different for the four HBV proteins and it could be easily understood since the functions are different for these proteins, the DNA polymerase undergoes highest selective pressure for drug resistance and the S protein have the highest pressure for immune escape. The X protein is the intracellular regulator for DNA replication and repair, transcription, and signal transduction. The core protein needs to be conserved for the structure stability of the virus. Different biological functions exert selective to these proteins and the mutation patterns for these HBV proteins are therefore different.

From the above analyses, we can conclude that the spectrum of HBV protein mutations have several trends which are different from the spectrum of disease causing amino acid mutations (Khan & Vihinen, 2007; see Fig. 7), (1) the serine and threonine have highest mutability and the tryptophan and methionine have lowest mutability, while the tryptophan was reported has highest mutability for disease causing mutations. (2) Most of the mutations are happened in hydrophobic and the polar amino acid groups. It can be seen from Fig. 8 that most of the mutations in HBV proteins did not change their physicochemical properties, especially the hydrophobicity, since the hydrophobic interaction is the main driven force for the protein structure. Figure 9 presented the detail of the mutations between the 20 amino acids. Most of the serine is mutated to alanine, proline, and valine. The threonine is frequently mutated to alanine, serine, and isoleucine. In addition, serine and proline are often mutated to each other.

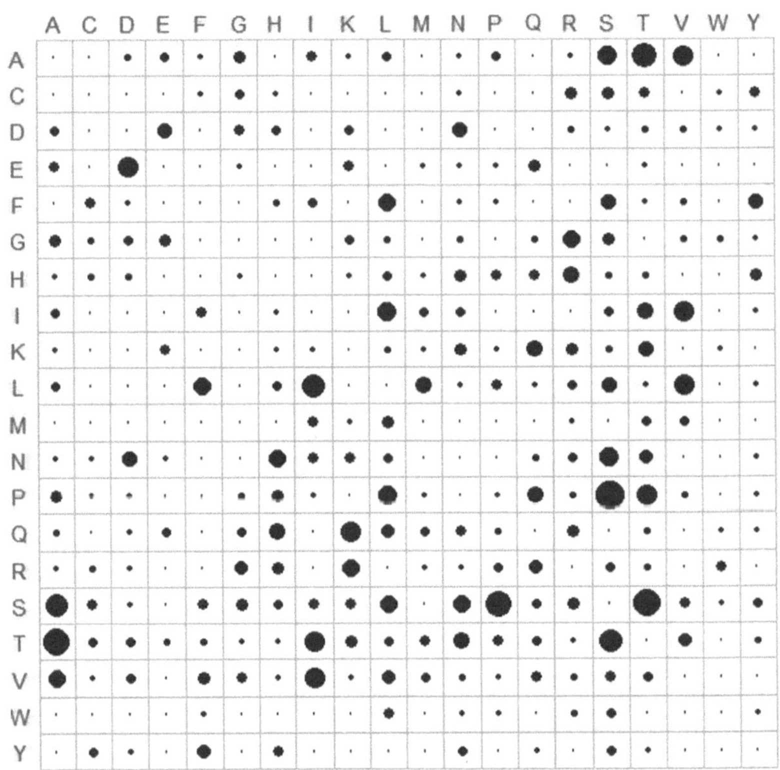

Figure 9 The mutation matrix circles for the 20-letter alphabet amino acids. The sizes of the circles correlate with the number of mutations.

3. BIOINFORMATICS ANALYSIS OF THE MUTATIONS

3.1 Mutations Happened in the Epitopic Regions

We take the sequences of the four HBV proteins as the inputs and applied the epitope prediction tools to predict the epitopic regions. We applied the online tool SYFPEITHI (http://www.syfpeithi.de/) to predict the MHC type 1 and type 2 epitopes (Rammensee, Bachmann, Emmerich, Bachor, & Stevanovic, 1999; Schuler, Nastke, & Stevanovikc, 2007), the online tool BepiPred (http://www.cbs.dtu.dk/services/BepiPred/) to predict the liner B cell epitopes (Larsen, Lund, & Nielsen, 2006). We then map the amino acid mutations to the predicted epitopes, we wish to know the hot spots for the mutation and identify the most active epitopes. The results are listed in Tables 2–5 for all MHC epitopes, Table 6 for the B cell epitopes of the four HBV proteins. Although further verification is needed, we marked the most active epitopes with bold fonts. It is obvious that most of the amino acid mutations happened in these epitopic regions.

Table 2 Mutation Distribution on Different MHC Epitopes of HBV DNA Polymerase

Epitope (9AA)	Start Position	MHC Type 1	Score	Number of Mutations
QHFRKLLLL	6	HLA-B*39:01	26	5
LEEELPRLA	21	HLA-B*41:01	24	1
DEGLNRRVA	30	HLA-B*45:01	22	3
RRVAEDLNL	35	HLA-B*14:02	27	2
AEDLNLGNL	38	HLA-B*40:01	23	5
VSIPWTHKV	48	HLA-B*15:16	29	1
TPSFPDIHL	76	HLA-B*07:02	25	4
KLIMPARFY	108	HLA-B*15:01	22	5
VTKYLPLDK	119	HLA-A*11:01	25	10
KYLPLDKGI	121	HLA-A*24:02	24	3
PLDKGIKPY	124	HLA-A*01	27	2
KPYYPEHVV	130	HLA-B*51:01	25	8
YPEHVVNHY	133	HLA-B*53:01	23	9

Continued

Table 2 Mutation Distribution on Different MHC Epitopes of HBV DNA Polymerase—cont'd

Epitope (9AA)	Start Position	MHC Type 1	Score	Number of Mutations
RETTRSASF	160	HLA-B*37	29	3
STRHGDESF	**189**	**HLA-B*57:01**	**18**	**10**
VEPSGSGHI	**257**	**HLA-B*49:01**	**23**	**13**
PHNTTESRL	380	HLA-B*15:10	22	1
TESRLVVDF	384	HLA-B*18	23	1
WPKFAVPNL	404	HLA-B*08	27	0
DVSAAFYHL	429	HLA-A*26	26	1
HPAAMPHLL	440	HLA-B*35:01	23	0
GLSRYVARL	453	HLA-A*02:01	28	1
LHDSCSRNL	478	HLA-B*38:01	24	6
RKIPMGVGL	513	HLA-B*39:02	26	0
SAICSVVRR	531	HLA-A*68:01	25	1
KSVQHLESL	558	HLA-B*58:02	26	3
PQEHIVQKI	607	HLA-B*13	26	8
NLYPVARQR	683	HLA-A*03	28	2
ADDPSRGRL	787	HLA-B*47:01	25	0
Epitope (15AA)	**Start Position**	**MHC Type 2**	**Score**	**Number of Mutations**
KLLLLDDEAGPLEEE	10	HLA-DRB1*0301 (DR17)	31	6
THKVGNFTGLYSSTV	53	HLA-DRB1*1501 (DR2b)	34	2
VGNFTGLYSSTVPVF	56	HLA-DRB1*0401 (DR4Dw4)	28	8
ARFYPNVTKYLPLDK	113	HLA-DRB1*1101	31	15
SFGVEPSGSGHIDNS	**254**	**HLA-DRB1*0701**	**30**	**35**
YTSFPWLLGCAANWI	757	HLA-DRB1*0101	34	2

Table 3 Mutation Distribution on Different MHC Epitopes of HBV S Protein

Epitope (9AA)	Start Position	MHC Type 1	Score	Number of Mutations
PRKGMGTNL	8	HLA-B*27:05	26	10
LSVPNPLGF	16	HLA-B*15:16	24	3
SVPNPLGFF	17	HLA-A*26	24	3
PDHQLDPAF	26	HLA-B*37	22	3
GILTTVPAA	**83**	**HLA-B*50:01**	**12**	**19**
RQSGRQPTP	99	HLA-B*41:01	14	5
QPTPISPPL	104	HLA-B*07:02	23	4
DTHPQAMQW	114	HLA-B*57:01	20	6
HQALQDPRV	128	HLA-B*13	22	8
LQDPRVRGL	131	HLA-B*39:02	20	5
DPRVRGLYF	133	HLA-B*08	27	5
RTGDPAPNM	167	HLA-B*58:02	23	10
DPAPNMENI	170	HLA-B*51:01	24	10
MENITSGFL	175	HLA-B*18	18	6
QAGFFLLTR	190	HLA-A*68:01	20	5
PPICPGYRW	240	HLA-B*53:01	24	5
RRFIIFLFI	252	HLA-B*27:09	26	3
LFILLLCLI	258	HLA-A*24:02	20	2
PLIPGSSTT	282	HLA-A*03	25	8
STTSTGPCK	**288**	**HLA-A*11:01**	**24**	**14**
LSVIWMMWY	336	HLA-A*01	20	7
VRFSWLSLL	342	HLA-B*14:02	28	4
FSWLSLLVP	344	HLA-B*45:01	18	4
LLVPFVQWF	349	HLA-B*15:01	20	6
ILSPFLPLL	382	HLA-A*02:01	20	9

Continued

Table 3 Mutation Distribution on Different MHC Epitopes of HBV S Protein—cont'd

Epitope (15AA)	Start Position	MHC Type 2	Score	Number of Mutations
APNMENITSGFLGPL	172	HLA-DRB1*0701	30	12
QAGFFLLTRILTIPQ	190	HLA-DRB1*0401 (DR4Dw4)	28	7
WTSLNFLGGTPVCLG	**210**	**HLA-DRB1*0101**	**36**	**15**
FLLVLLDYQGMLPVC	267	HLA-DRB1*0301 (DR17)	37	12
LLVLLDYQGMLPVCP	268	HLA-DRB1*1501 (DR2b)	34	14

Table 4 Mutation Distribution on Different MHC Epitopes of HBV X Protein

Epitope (9AA)	Start Position	MHC Type 1	Score	Number of Mutations
ARLCCQLDP	3	HLA-B*14:02	20	6
CQLDPARDV	7	HLA-B*13	22	4
DPARDVLCL	10	HLA-B*07:02	24	4
VGAESRGRP	21	HLA-B*58:02	22	6
PLSGPLGTL	29	HLA-A*02:01	26	16
TLPSPSPSA	**36**	**HLA-B*50:01**	**12**	**22**
VPADHGAHL	45	HLA-A*24:02	15	8
SAGPCALRF	65	HLA-B*53:01	17	2
LRFTSARRM	71	HLA-B*27:05	25	4
RRMETTVNA	77	HLA-B*41:01	15	7
ETTVNAHQV	80	HLA-B*15:16	22	19
NAHQVLPKV	84	HLA-B*51:01	24	19
AHQVLPKVL	85	HLA-B*15:10	24	19
QVLPKVLHK	87	HLA-A*03	31	17
PKVLHKRTL	90	HLA-B*39:02	21	6

Table 4 Mutation Distribution on Different MHC Epitopes of HBV X Protein—cont'd

Epitope (9AA)	Start Position	MHC Type 1	Score	Number of Mutations
MSTTDLEAY	103	HLA-A*01	19	7
STTDLEAYF	104	HLA-B*57:01	18	7
EELGEEIRL	121	HLA-B*40:01	26	10
GEEIRLKVF	124	HLA-B*37	26	14
EIRLKVFVL	126	HLA-A*26	27	13
RLKVFVLGG	128	HLA-B*15:01	14	6
Epitope (15AA)	**Start Position**	**MHC Type 2**	**Score**	**Number of Mutations**
ARDVLCLRPVGAESR	12	HLA-DRB1*0101	32	6
LRPVGAESRGRPLSG	18	HLA-DRB1*1101	22	12
PSAVPADHGAHLSLR	42	HLA-DRB1*0301 (DR17)	29	18
VCAFSSAGPCALRFT	60	HLA-DRB1*0701	24	4
ALRFTSARRMETTVN	70	HLA-DRB1*0401 (DR4Dw4)	28	7

Table 5 Mutation Distribution on Different MHC Epitopes of HBV Core Protein

Epitope (9AA)	Start Position	MHC Type 1	Score	Number of Mutations
MQLFHLCLI	1	HLA-B*13	21	2
CPTVQASKL	14	HLA-B*35:01	21	2
KEFGASVEL	36	HLA-B*40:01	29	2
LSFLPSDFF	45	HLA-B*57:01	18	3
DFFPSIRDL	51	HLA-B*15:10	15	2
FFPSIRDLL	52	HLA-A*24:02	21	2
RDLLDTASA	57	HLA-B*47:01	18	2
LLDTASALY	59	HLA-A*01	28	3
ALYREALES	65	HLA-A*03	26	3

Continued

Table 5 Mutation Distribution on Different MHC Epitopes of HBV Core Protein—cont'd

Epitope (9AA)	Start Position	MHC Type 1	Score	Number of Mutations
LESPEHCSP	71	HLA-B*45:01	22	5
HHTALRQAI	80	HLA-B*38:01	19	3
LEDPASREL	105	HLA-B*37	22	9
DPASRELVV	107	HLA-B*51:01	26	8
ASRELVVSY	109	HLA-B*58:02	23	9
YVNVNMGLK	117	HLA-A*11:01	23	4
NVNMGLKIR	119	HLA-A*68:01	22	6
GLKIRQLLW	123	HLA-B*08	25	4
LLWFHISCL	129	HLA-A*02:01	26	1
GRETVLEYL	140	HLA-B*27:09	23	6
RETVLEYLV	141	HLA-B*49:01	20	6
TVLEYLVSF	143	HLA-A*26	26	3
PPNAPILST	163	HLA-B*07:02	22	2
STLPETTVV	170	HLA-B*15:16	27	3
RRRGRSPRR	179	HLA-B*27:05	30	4
Epitope (15AA)	**Start Position**	**MHC Type 2**	**Score**	**Number of Mutations**
LWGMDIDPYKEFGAS	27	HLA-DRB1*0301 (DR17)	29	4
IRDLLDTASALYREA	56	HLA-DRB1*0701	30	6
ILCWGELMNLATWVG	**88**	**HLA-DRB1*0401 (DR4Dw4)**	**28**	**10**
RELVVSYVNVNMGLK	111	HLA-DRB1*1501 (DR2b)	34	9
TLPETTVVRRRGRSP	171	HLA-DRB1*1101	27	8
KLCLGWLWGMDIDPY	21	HLA-DRB1*0101	30	3

Table 6 Mutation Distribution in B Cell Epitopes of HBV Core Protein

Proteins	Epitopes	Number Of Mutations	Protein	Epitopes	Number Of Mutations
Core protein	16–17	1	DNA polymerase	128–135	1
	32–39	1		161–178	8
	71–79	5		190–199	19
	103–110	12		204–213	9
	156–169	6		224–243	35
	177–212	11		**248–273**	**57**
X protein	22–50	42		291–300	16
	65–66	1		306–319	24
	68	0		331–334	3
	78	3		350–357	6
	105	2		364–369	1
	122–123	1		379–385	1
S protein	1–19	20		395–403	9
	21	0		407	0
	23–25	0		473	1
	27–83	**46**		582–588	4
	87–122	25		604–607	2
	132–134	2		625	0
	136	0		627	0
	142–161	24		647–649	0
	166–177	14		668	1
	205	0		689–692	0
	207–208	0		700–706	0
	221–243	25		740	0
	284–307	**46**		744–746	2
	313–321	20		783–793	0
DNA polymerase	17–28	1		807–813	3
	30–34	2		819–827	2
	67–80	15		841–843	1
	100–102	2			

Table 7 HBV Protein Mutations Distributed in Different Epitopes or Nonepitope Regions

HBV Proteins	Conservation	MHC1	MHC2	B Cell	All Epitope	Nonepitope	All Mutation
DNA polymerase	Neutral	90	64	220	227	287	514
	Deleterious	4	2	8	2	10	12
S protein	Neutral	99	30	151	213	57	269
	Deleterious	32	18	72	92	11	103
X protein	Neutral	89	36	42	14	90	104
	Deleterious	20	7	7	6	21	27
Core protein	Neutral	52	33	22	75	2	77
	Deleterious	6	4	4	10	0	10

3.2 Evolutionary or Conservation Analysis of the Mutations

To study the relationship between the evolutionary pressure and the conservation, we further analyzed the conservations of all the mutations, we used the online tool SIFT (http://sift.bii.a-star.edu.sg/) to classify the mutations as neutral and deleterious, the results are listed in Table 7, as expected, the mutations happened in epitopic regions are less conserved than those in the nonepitopic regions, this indicate that most of the mutations in the epitopic regions undergo higher selective pressures than those in the nonepitopic regions.

4. DISCUSSION AND CONCLUSION

With the environment changes, nowadays more new viruses are generated and their evolution rates are high and we face challenges to overcome the attacks of these virus. To understand the complex interactions between host–virus–environments is very important to design drugs or vaccines rationally for the antivirus battle, the amino acid mutations in virus proteins is an important resource to investigate the functions and the evolution of virus replication and infection, the first hand data is also necessary to the future modeling and simulation the complex systems. We need to distinguish the neutral mutations from the deleterious mutations and even identify the key or driver mutations for the rational drug and vaccine design. Our next step is to reason the mutation patterns we found in this study. We will extend our research to other viruses to identify popular patterns in the viral protein mutations.

In this study, we collected all the reported amino acid mutations in the four HBV proteins, we identified several important patterns for the amino acid mutations, and we also located the most active MHC and B cell epitopes. These results provide important information for the next step investigation of amino acid mutation in virus proteins.

ACKNOWLEDGMENTS

We gratefully acknowledge financial support from the National Natural Science Foundation of China grants (31470821, 91230117, and 31170795).

Competing interest: The authors declare that they have no competing interests.

REFERENCES

Altuvia, Y., & Margalit, H. (2004). A structure-based approach for prediction of MHC-binding peptides. *Methods*, *34*(4), 454–459.

Bhasin, M., & Raghava, G. P. (2003). Prediction of promiscuous and high-affinity mutated MHC binders. *Hybridoma and Hybridomics*, *22*(4), 229–234.

Bordner, A. J., & Abagyan, R. (2006). Ab initio prediction of peptide-MHC binding geometry for diverse class I MHC allotypes. *Proteins*, *63*(3), 512–526.

Chen, C. M. (2006). CiteSpace II: Detecting and visualizing emerging trends and transient patterns in scientific literature. *Journal of the American Society for Information Science and Technology*, *57*(3), 359–377.

Chen, J., & Shen, B. (2009). Computational analysis of amino acid mutation: A proteome wide perspective. *Current Proteomics*, *6*(4), 228–234.

Chen, J., Wang, Y., Guo, D., & Shen, B. (2012). A systems biology perspective on rational design of peptide vaccine against virus infections. *Current Topics in Medicinal Chemistry*, *12*(12), 1310–1319.

Chen, C. J., Yang, H. I., Su, J., Jen, C. L., You, S. L., Lu, S. N., et al. (2006). Risk of hepatocellular carcinoma across a biological gradient of serum hepatitis B virus DNA level. *Jama-Journal of the American Medical Association*, *295*(1), 65–73.

Cui, J., Han, L. Y., Lin, H. H., Tang, Z. Q., Jiang, L., Cao, Z. W., et al. (2006). MHC-BPS: MHC-binder prediction server for identifying peptides of flexible lengths from sequence-derived physicochemical properties. *Immunogenetics*, *58*(8), 607–613.

Donnes, P., & Kohlbacher, O. (2006). SVMHC: a server for prediction of MHC-binding peptides. *Nucleic Acids Research*, *34*, W194–197 (Web Server issue).

Doytchinova, I. A., & Flower, D. R. (2006). Class I T-cell epitope prediction: Improvements using a combination of proteasome cleavage, TAP affinity, and MHC binding. *Molecular Immunology*, *43*(13), 2037–2044.

Du, X., Dong, L., Lan, Y., Peng, Y., Wu, A., Zhang, Y., et al. (2012). Mapping of H3N2 influenza antigenic evolution in China reveals a strategy for vaccine strain recommendation. *Nature Communications*, *3*, 709.

El-Manzalawy, Y., & Honavar, V. (2010). Recent advances in B-cell epitope prediction methods. *Immunome Research*, *6*(Suppl 2), S2.

Goodswen, S. J., Kennedy, P. J., & Ellis, J. T. (2014). Enhancing in silico protein-based vaccine discovery for eukaryotic pathogens using predicted peptide-MHC binding and peptide conservation scores. *PLoS One*, *9*(12), e115745.

Greenbaum, J. A., Andersen, P. H., Blythe, M., Bui, H. H., Cachau, R. E., Crowe, J., et al. (2007). Towards a consensus on datasets and evaluation metrics for developing B-cell epitope prediction tools. *Journal of Molecular Recognition*, *20*(2), 75–82.

Guan, P., Doytchinova, I. A., Zygouri, C., & Flower, D. R. (2003). MHCPred: A server for quantitative prediction of peptide-MHC binding. *Nucleic Acids Research*, *31*(13), 3621–3624.

Hammer, J. (1995). New methods to predict MHC-binding sequences within protein antigens. *Current Opinion in Immunology*, *7*(2), 263–269.

Hayer, J., Jadeau, F., Deleage, G., Kay, A., Zoulim, F., & Combet, C. (2013). HBVdb: A knowledge database for Hepatitis B Virus. *Nucleic Acids Research*, *41*(Database issue), D566–570.

Huang, J., Deng, Q., Wang, Q., Li, K. Y., Dai, J. H., Li, N., et al. (2012). Exome sequencing of hepatitis B virus-associated hepatocellular carcinoma. *Nature Genetics*, *44*(10), 1117–1121.

Jeng, W.-J., Sheen, I. S., & Liaw, Y.-F. (2010). Hepatitis B virus DNA level predicts hepatic decompensation in patients with acute exacerbation of chronic hepatitis B. *Clinical Gastroenterology and Hepatology*, *8*(6), 541–545.

Kamili, S., Sozzi, V., Thompson, G., Campbell, K., Walker, C. M., Locarnini, S., et al. (2009). Efficacy of hepatitis B vaccine against antiviral drug-resistant hepatitis B virus mutants in the chimpanzee model. *Hepatology*, *49*(5), 1483–1491.

Khan, S., & Vihinen, M. (2007). Spectrum of disease-causing mutations in protein secondary structures. *BMC Structural Biology*, 7, 56.

Kim, H., Oh, B. K., Roncalli, M., Park, C., Yoon, S. M., Yoo, J. E., et al. (2009). Large liver cell change in hepatitis B virus-related liver cirrhosis. *Hepatology*, *50*(3), 752–762.

Kringelum, J. V., Lundegaard, C., Lund, O., & Nielsen, M. (2012). Reliable B cell epitope predictions: impacts of method development and improved benchmarking. *PLoS Computational Biology*, *8*(12), e1002829.

Larsen, J. E., Lund, O., & Nielsen, M. (2006). Improved method for predicting linear B-cell epitopes. *Immunome Research*, *2*, 2.

Lian, Y., Ge, M., & Pan, X. (2014). EPMLR: Sequence-based linear B-cell epitope prediction method using multiple linear regression. *BMC Bioinformatics*, *15*(1), 414.

Liaw, Y. F., & Chu, C. M. (2009). Hepatitis B virus infection. *Lancet*, *373*(9663), 582–592.

Molero-Abraham, M., Lafuente, E. M., & Reche, P. (2014). Customized predictions of peptide-MHC binding and T-cell epitopes using EPIMHC. *Methods in Molecular Biology*, *1184*, 319–332.

Osiowy, C., Giles, E., Tanaka, Y., Mizokami, M., & Minuk, G. Y. (2006). Molecular evolution of hepatitis B virus over 25 years. *Journal of Virology*, *80*(21), 10307–10314.

Rammensee, H., Bachmann, J., Emmerich, N. P., Bachor, O. A., & Stevanovic, S. (1999). SYFPEITHI: Database for MHC ligands and peptide motifs. *Immunogenetics*, *50*(3–4), 213–219.

Rhee, S. Y., Margeridon-Thermet, S., Nguyen, M. H., Liu, T. F., Kagan, R. M., Beggel, B., et al. (2010). Hepatitis B virus reverse transcriptase sequence variant database for sequence analysis and mutation discovery. *Antiviral Research*, *88*(3), 269–275.

Schuler, M. M., Nastke, M. D., & Stevanovikc, S. (2007). SYFPEITHI: Database for searching and T-cell epitope prediction. *Methods in Molecular Biology*, *409*, 75–93.

Shen, B., & Vihinen, M. (2004). Conservation and covariance in PH domain sequences: Physicochemical profile and information theoretical analysis of XLA-causing mutations in the Btk PH domain. *Protein Engineering Design and Selection*, *17*(3), 267–276.

Shen, W. J., Wei, Y. T., Guo, X., Smale, S., Wong, H. S., & Li, S. C. (2014). MHC binding prediction with KernelRLSpan and its variations. *Journal of Immunological Methods*, *406*, 10–20.

Shepard, C. W., Simard, E. P., Finelli, L., Flore, A. E., & Bell, B. P. (2006). Hepatitis B virus infection: Epidemiology and vaccination. *Epidemiologic Reviews*, *28*, 112–125.

Singh, H., Ansari, H. R., & Raghava, G. P. (2013). Improved method for linear B-cell epitope prediction using antigen's primary sequence. *PLoS One*, *8*(5), e62216.

Sweredoski, M. J., & Baldi, P. (2008). PEPITO: Improved discontinuous B-cell epitope prediction using multiple distance thresholds and half sphere exposure. *Bioinformatics*, *24*(12), 1459–1460.

Vita, R., Overton, J. A., Greenbaum, J. A., Ponomarenko, J., Clark, J. D., Cantrell, J. R., et al. (2015). The immune epitope database (IEDB) 3.0. *Nucleic Acids Research*, *43*(Database issue), D405–412.

Wan, J., Liu, W., Xu, Q., Ren, Y., Flower, D. R., & Li, T. (2006). SVRMHC prediction server for MHC-binding peptides. *BMC Bioinformatics*, 7, 463.

Yao, B., Zheng, D., Liang, S., & Zhang, C. (2013). Conformational B-cell epitope prediction on antigen protein structures: A review of current algorithms and comparison with common binding site prediction methods. *PLoS One*, *8*(4), e62249.

Zhang, Q., Wang, P., Kim, Y., Haste-Andersen, P., Beaver, J., Bourne, P. E., et al. (2008). Immune epitope database analysis resource (IEDB-AR). *Nucleic Acids Research*, *36*, W513–518 (Web Server issue).

Zhao, L., Wong, L., Lu, L., Hoi, S. C., & Li, J. (2012). B-cell epitope prediction through a graph model. *BMC Bioinformatics*, *13*(Suppl 17), S20.

Zhou, H. Y., Luo, Y., Chen, W. D., & Gong, G. Z. (2015). HBV mutation may play a role in HCC recurrence: a systematic review and meta-regression analysis. *Journal of Gastroenterology and Hepatology*.

AUTHOR INDEX

Note: Page numbers followed by "*np*" indicate footnotes.

C

F

G

H

I

J

K

M

Q

R

S

W

X

Y

Z

SUBJECT INDEX

Note: Page numbers followed by "*f*" indicate figures and "*t*" indicate tables.

T

U

V

W

X

Kolibri 360 GmbH
Ferdinand-Jühlke-Straße 7
99095 Erfurt, Deutschland
produktsicherheit@kolibri360.de